The Car Design Yearbook 7

Stephen Newbury
and Tony Lewin

The Car Design Yearbook

the definitive annual guide to all new concept
and production cars worldwide

MERRELL
LONDON · NEW YORK

Contents

Trends, Highlights, Predictions

Trends, Highlights, Predictions

In each of the seven years since we began compiling *The Car Design Yearbook* we have been able to highlight the major new models of the season and identify the broader trends at work shaping the car industry and its products. Last year, we saw how German premium carmakers had at last begun to embrace hybrid technology; this year marks an intensification of that trend and its expansion into unexpected areas. Of all the trends apparent in the past twelve months, perhaps the most significant is the return of the battery-powered car. Once dismissed as too heavy, too slow, and too costly, it is now firmly back on the agenda—whether as a pure electric, or in combination with a small range-extending motor or fuel cell. The buzzword on every show floor was "plug-in," the magic connection that could enable most everyday commuting to be done on electricity alone.

Such vehicles blur the boundary between what is an electric vehicle and what is a hybrid; unquestionably, however, what we are looking at is a major step change in powertrain technology. This is not just scientists in white coats speaking: equally clear is the much greater emphasis by manufacturers in pushing environmental technology to consumers.

The threat of very much tighter CO_2 emissions standards in Europe and North America has focused the minds of the world's automotive engineers as never before. This has been a wake-up call to both consumers and manufacturers, and awareness of the need to consider the global climate change picture is at an all-time high. The current growth in hybrid vehicles is just the beginning. Already, half a dozen car manufacturers offer hybrid versions of some of their models; by 2012, every major automaker will include at least one hybrid line, with major firms, such as Toyota, committed to providing a hybrid option on all their major models.

Consumers are now asking retailers in all sectors for ethical and environmentally responsible products. In years to come we can expect to see a firming-up of these values in

The Citroën C-Cactus breaks with show-car tradition by proposing a simple, back-to-basics concept. The design offers light weight and minimal equipment, the cost savings being channeled into an efficient diesel hybrid engine. Novel design solutions include a fixed-hub steering wheel and the elimination of the dashboard.

personal transportation, too: business executives driving gas-guzzling luxury cars could be made to feel very uncomfortable for doing so. In the UK, for instance, government taxes on the highest-CO_2 cars have increased, and the only real protests have come from a single manufacturer, of high-emitting sports cars and SUVs: Porsche.

Recycling should not be confined to the home. Industry has a responsibility to produce goods from recycled materials and goods that themselves can be easily recycled. The Citroën C-Cactus concept is one such example that reduces the environmental impact of road travel. It is made from far fewer components than an ordinary car: in particular, the interior trim is scaled right back to a simple functional specification with few frills. Air-conditioning is standard, but the windows do not open fully.

If 2007 was the year that China dominated the auto industry's dealings, 2008 has been the

year when Russia and India snatched the spotlight. The growing wealth of the Russian economy through the sale of oil and gas has unleashed a spectacular boom in car sales, providing much-needed relief for carmakers hit by a stagnant European market and declining North American sales. Luxury vehicles have led the way, but now car producers from Europe, the USA, Japan, and Korea are rushing to set up shop on Russian soil to take advantage of the middle-class car-buying spree that is about to follow.

India, too, is much in the news, as much for its own internal potential as for the overseas acquisitions of its major industrial groups. Prime among these is of course Tata Motors and its US$2.3 billion deal to purchase Jaguar Land Rover from Ford, ensuring a secure future for these revered brands.

Yet Tata would have been in the headlines even had it not chosen to bid for the British

Opposite
The Jaguar XF marks a return to form of Jaguar style after a prolonged period of retro designs. The interior has been particularly highly praised; its simplicity is likely to be influential for some time to come.

Above
Ford Group design director Peter Horbury sees the Lincoln MKT as a Learjet for the road—a luxury grand tourer that brings together elements from sports coupés, upscale limousines, and crossover utilities.

makers: its Nano, the world's cheapest car at just US$2500, attracted immense global interest on its unveiling in New Delhi in January 2008. So sensational was this vehicle that it was the talk of the Detroit show the following week, even though it was not actually present. In one of this year's feature articles we take a detailed look at the Nano and explore just how Tata has managed to make a car for such a low retail price. The Nano showcases India's clever design thinking, and we show that it is just as tough a task to design for low cost as it is to engineer an expensive luxury vehicle, such as a high-specification Mercedes. The reality of commercial restriction inspires real innovation, and we expect that India's pool of well-educated, highly motivated engineers, who can work at considerably lower cost than their colleagues in Western economies, will become a driving force in vehicle engineering in decades to come. And as the Indian market comes to demand more sophisticated vehicles, we can expect to see more technology partnerships with Western

Opposite and above
The Fiat 500 is an eagerly awaited revival of a 1950s icon and incorporates elegant reinterpretations of period details, such as the cream interior detailing.

Top
The latest Audi A4 is a skillful update that succeeds in maintaining close touch with Audi identity. Shapely LED strings in the headlights are set to become a powerful brand signature.

companies that will license technology and even share manufacturing with Indian producers.

Trends in aesthetic design are also becoming clear. The drive to strengthen the individual visual identity of cars continues unabated. Bold, distinctive grilles in particular are the main thrust, some notable examples being the Chrysler ecoVoyager and Lincoln MKT concepts and the Jaguar XF.

LED lights are now being used more extensively on premium cars. As an example, the new Audi A4 has LED running lights incorporated into the headlamps. These lights are being used

as an interesting graphical shape in the design, not just as a practical device; indeed, different Audis can be readily distinguished by the illuminated signatures of their LED running lights.

Strong interest in retro-inspired vehicles remains. The Fiat 500 is perhaps the best example: relaunched in 2007, fifty years after the birth of the original, this cute car still captures the hearts of many. What is more, the very fuel-efficient motoring it provides is as relevant today as it was back in the 1950s. Like the new Mini, the new car is bigger, much more refined, and far safer than the original.

Above
The appealing Daihatsu Mud Master light 4x4 leisure vehicle is one of the more practical concepts to have appeared at the Tokyo show.

Opposite
Now a second-generation concept, the Nissan Pivo continues to fascinate. The spherical cabin can rotate on the four-wheeled chassis, eliminating the need for reversing, and the chassis can move like a crab sideways into a parking space.

From the Tokyo show we expect a number of wacky vehicles, and the late-2007 show certainly did not disappoint. The Daihatsu Mud Master is a compact adventure vehicle with some off-road capability; the space-age Honda Puyo, with its bubble roof, seemed to have more potential in a zero-emission zone. Nissan's Pivo 2 is the Japanese answer to future urban transport and drew much attention. This product-designed vehicle has pivoting wheels and a revolving bubble cabin for the ultimate in maneuverability.

As with every edition of *The Car Design Yearbook*, we feature a couple of established car

designers who are influential in the industry today. We chart the story of how they rose to become leaders in their field, the models for which they have been responsible and those that have shaped their careers, and what challenges they are facing today. In this edition we look at the careers of Simon Cox, GM's head of advanced design, based in the UK, and Keith Ryder, head of design at Peugeot in France.

Future vehicle innovation will be driven by environmental considerations. These, in turn, are shaped by legislators, often in consultation with auto-industry experts. It will not be only fuel

economy on the agenda, but also stronger measures of sustainability as well as control of the environmental costs of manufacturing and distribution. Today's policymakers could be about to set standards that will force the biggest shift in vehicle design and engineering for decades. And it will be up to the many talents in car-design studios, from California to Stuttgart and from Barcelona to Bombay, to develop new design languages to communicate the environmentally responsible vehicle message in the showroom and ensure that customers continue to be passionate about their cars.

Vehicle Sensing Technologies

The Tata Nano

Vehicle Sensing Technologies

The current buzzword (or phrase) among automotive engineers—especially electronics specialists—is "data fusion." Translated into everyday English, this means the combining of information from a variety of sources (such as cameras, sensors, and scanners) to enable a vehicle to know what is going on around and even inside it. This, in turn, allows the vehicle to respond appropriately to its surroundings—in other words, to become proactive and "intelligent," rather than remaining a simple, passive machine. This is the real new frontier in car design.

In truth, cars have been growing steadily more intelligent ever since electronics first found their way under the hood and behind the dashboard in the 1960s. We have long been accustomed to interior lights that fade gently once you have closed the door, wipers that go on automatically as soon as it rains, lights that illuminate by themselves when it gets dark, and stereo systems that play louder as speed rises. More recently, parking-distance sensors, reversing cameras, and even automatic parking sequences have helped to simplify driving in town. Many models have systems that buzz or beep if you stray out of lane; Volvos now warn of cars or motorcycles lurking in the blind spot between the three rearview mirrors; and satellite navigation setups can find the best route from A to B without any effort on the driver's part.

All these are undoubtedly intelligent developments, ideas that give a positive push to comfort and safety. Yet, compared with what is about to come, these existing, isolated systems will be seen as simple, primitive, and limited in their effectiveness. For by networking together the many different sensor systems soon to be built into the vehicle, the car will be able to make informed judgments and react intelligently rather than as a mere programmed reflex. The benefit is more than the sum of the individual parts.

Take radar systems, for example. These have

Opposite
Mercedes' alert system uses vehicle-to-vehicle and infrastructure-to-vehicle communications to warn of obstructions ahead.

Above
BMW and Google pioneered a system that allows a driver to plot a route in Google Maps on the home or office computer, and then beam it directly into the car's navigation system.

Above, right
Volvo's BLIS system scans rearward to detect vehicles or objects in the driver's blind spots; the presence of an object is indicated with a warning triangle in the mirror.

been available on several types of luxury car for a few years, and are typically used to allow the car to maintain a constant distance or time-gap from the vehicle in front, irrespective of how much that car's speed varies. That same radar can quite easily provide the information needed to calculate how quickly the vehicle is approaching another car or object, and from there it is a relatively simple step to predicting the likelihood of a collision. That information, in turn, can be used to trigger an alarm, to apply the brakes, or, if a collision is judged unavoidable, to prime such safety systems as airbags and seatbelts so that they are in the optimum position.

Such systems, known as adaptive cruise control and collision-mitigation braking, are beginning to filter down into more affordable cars, notably several Honda and Ford models. Yet, impressive though such features as automatic emergency braking undoubtedly are, there is a good deal more still to come.

Europe, North America, and Japan are all committed to ambitious targets to reduce the number of people killed or injured in or by cars. Major casualty reductions have already been achieved as vehicles have become better designed and drivers and passengers remember to use their safety features. But as the protection

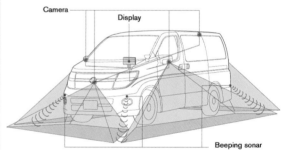

Camera

Display

Beeping sonar

Top
Bosch is one of several suppliers to have developed systems that scan the road ahead and apply brakes if a collision appears to be imminent and the driver does not respond. The speed reduction reduces the severity of the accident.

Above
Nissan's all-round sensing system is designed to aid low-speed maneuvering of big vehicles and to prevent accidents with pedestrians and cyclists.

Left
This prototype of a pedestrian-recognition system in a BMW demonstrator car can distinguish pedestrians from surrounding objects and warn the driver of the risk of an accident.

Opposite
Volvo City Safety, now installed in the XC60 SUV, is a pioneering concept allowing automated braking in urban conditions, helping to prevent whiplash injuries and reduce damage to vehicles.

offered by the vehicle itself begins to approach a very high ceiling, further vehicle-based improvement is ever harder to find. Attention is thus starting to shift to the most unpredictable element of all: the behavior of the driver.

Already, antilock brakes prevent a heavy-footed driver from locking the wheels and skidding on a slippery road; electronic stability control, again entirely automatic, goes a step further in manipulating individual brakes to keep the car on its intended course; and even so relatively modest a car as the Peugeot 207 RC has a system that intervenes in its steering to keep the car straight under emergency braking. These flatter a driver's performance. Now, monitoring, surveillance, and scanning systems are beginning to protect us against normal human weaknesses, such as becoming drowsy on a long journey, being distracted when, say, inserting a CD into the stereo, or failing to notice another road user.

Lexus and Volvo both offer mechanisms that monitor the driver's face and eye movements to ascertain that he or she is remaining alert; in

the Lexus, the car will brake automatically if the driver fails to notice an obstacle. The Lexus looks backward, too: rearward-facing radars track vehicles coming up fast behind, and if they are on a collision course the system will swiftly adjust the seats and head restraints so as to minimize whiplash injury risk to the passengers on board.

The logical extension of this is for the car to put out a 360-degree sensory field to monitor its entire surroundings, in much the same way as an air traffic controller's radar screen gives a complete picture of aircraft positions, courses, and speeds in each zone of control. In fact, this already exists in a simple form, in the shape of low-speed parking sensors that, on certain models, such as BMWs and Nissans, are able to provide a bird's-eye view of the car and surrounding obstacles to aid parking or maneuvering. However, the task of upgrading this walking-pace thinking to full higher-speed collision avoidance is a daunting one, riddled as it is with both technical and legal challenges.

Carmakers and equipment suppliers have invested heavily in powerful onboard computer

systems capable of combining and interpreting the data flowing in from the many cameras and sensors. One of the principal challenges is to identify and classify all the objects picked up by the various scanning systems: as an illustration of the difficulties involved, a soft-drink can lying in the road can give just as strong a radar signal as a complete vehicle.

Even once object classification is fully mastered, a still bigger and potentially more controversial step remains. The vehicle may now know everything that is going on around it, but how should that information be used?

Today's cars have taken a cautious step forward in using radar information to trigger warnings or even apply limited emergency braking when a collision is imminent. Volvo has been bolder with its newly introduced City Safety on the XC60. This, as its name implies, uses a forward-facing laser sensor to scan the road up to 33 feet ahead: active at speeds of up to 20 mph, it will apply the brakes if the driver fails to notice a stationary or slowing vehicle ahead.

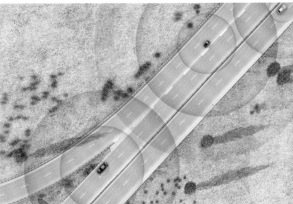

Legal concerns have deterred carmakers from implementing full automatic braking, which could prevent many accidents rather than just lessening their severity. This is largely because regulations stipulate that the vehicle must always be under the driver's direct control, but perhaps also because carmakers fear the consequences of a malfunction. Engineers agree, too, that it would be technically possible for the computer to intervene if an accident looked likely, and steer as well as brake the vehicle to avoid the accident completely. Yet most companies refuse to contemplate such automated actions, for fear that in steering to avoid one accident the vehicle might cause a worse one elsewhere.

Less controversial are systems in which the vehicle communicates with the road environment (for example, recognizing traffic signs) or with satellite navigation services. Some suppliers have even demonstrated systems by which cars communicate with one another to warn of such hazards as a stranded vehicle, an icy stretch of road, or a traffic jam.

Infrastructure-to-car communications could see important warnings given to drivers, or even speed limits automatically enforced as the vehicle passes the restriction signs. BMW is investigating mechanisms for preventing motorists from driving onto highways the wrong way (a regular problem in Germany); electronics supplier Continental is promoting similar schemes that would flag a broken-down vehicle or enable an emergency-service vehicle to trigger successive sets of traffic lights to allow it a clear run to its destination. E-Call, a further variation on this theme, sends out an automatic distress call in the event of an accident severe enough to prompt airbag deployment; it is likely to become a legal requirement in Europe before long.

The extensive mapping database held by satellite navigation systems opens up further possibilities. The mapping and positioning are sufficiently accurate that the system can warn the driver of a narrow bridge or a sharp bend immediately ahead; it can even recommend a suitable speed and perhaps also apply the

Above, left
One of the safety systems under investigation at BMW is designed to prevent motorists entering a highway in the wrong direction. This is a frequent cause of accidents on Germany's autobahns.

Above, top and bottom
Mercedes-Benz and other carmakers are studying the potential of automated car-to-car signaling (above) to reduce accidents; another opportunity is active road signs (top), which could trigger an alert on the dashboard and even impose a speed limit on the vehicle.

brakes if the driver is traveling too fast. Already, systems marketed by Nissan and Toyota in Japan are capable of this, and in a further extension topographical data is superimposed to give the car and the driver information on hills and gradients. This can in turn be used to improve efficiency and save fuel; transmission supplier ZF has developed a bus transmission that can be programmed to give the best efficiency on a range of bus routes in a city.

A further and perhaps more far-fetched application of sat-nav data is to improve comfort. Japanese electronics specialist Aisin is working on an interactive suspension system that interfaces with the navigation system to recall where the bumps in the road are: when the vehicle next travels that road, the suspension is able to anticipate each bump and thus improve comfort. If the road has been repaired in the interim, the system's memory is updated.

Nissan, working with Matsushita on the cold northern Japanese island of Hokkaido, is running a field trial of 100 customer cars fitted with a slip-hazard warning system. Real-time information from the Carwings navigation system combines with a historical database of skid-accident locations and automated feedback from other vehicles to warn the driver where there is a risk of ice; each time a vehicle's ABS system is triggered, for example, indicating the possibility of slippery conditions, the car's position and speed and other information are added to the database.

There are some who worry that such top-down monitoring of cars from satellites smacks of Big Brother and centralized control of individual movement: indeed, it would be perfectly possible to track many millions of vehicles for almost any purpose, be it marketing or advertising, law enforcement or road charging. But with the European Commission committed to halving traffic casualties and excess speed being one of the major causes of accidents, satellite speed monitoring or even speed-limit imposition may come to be seen as a politically acceptable reality, just like closed-circuit cameras in our buildings and city centers.

Forward-looking camera and radar systems draw in a tremendous amount of information. The current challenge for these systems is reliably to recognize and classify the objects being detected and to decide on the appropriate response, all within a fraction of a second.

The Tata Nano

It is not often that the launch of a new car sparks such global interest, especially when it is one designed to appeal to some of the world's least affluent customers. Yet when *Time* magazine recently listed a number of the most important cars in history, it included the tiny Tata Nano in its selection.

When Tata unveiled the Nano in January 2008 there was real astonishment, not so much at the design or the concept, but at the selling price: when it goes on sale in late 2008, Tata's "people's car" will have a selling price of 100,000 (known as 1 lakh) rupees (Rs). That is just US$2500—less than the price of some of the luxury options on a midsize European car.

For those who know little about the company, Tata Motors is a giant; in fact it is India's largest automotive company, with revenues of US$7.2 billion in 2006–07. With more than four million Tata vehicles plying the roads and tracks of India, it is the leader in commercial vehicles and the second-largest in passenger vehicles. It is also the world's fifth-largest medium and heavy truck manufacturer and the second-largest heavy bus manufacturer.

So why should such a prosperous conglomerate as Tata bother trying to manufacture such a low-cost car? After all, small, cheap cars are notorious in the car business for delivering minuscule or even negative profits.

The explanation lies in Tata managing director Ratan Tata's passionately held belief that India needs safer transport for its families—a realization that came to him one day when a motorcycle carrying a family of five pulled up next to him in a line of traffic. Not long afterward, he publicly committed Tata to building a four-seater car priced at Rs 1 lakh—halfway between the cost of a motorcycle and a conventional car. Industry analysts and other automakers were skeptical as to whether the job could be done: several speculated that the vehicle would have to be no more than a four-wheeled motorcycle or a tent on wheels to sell at so low a price.

Unusually, the shape, specification, and configuration of the 1 lakh car remained a complete secret until, under the gaze of the world's media at the January 2008 New Delhi show, the Nano was unveiled as a modern, smoothly styled four-seater (see the review on page 250). The specification sheet revealed that it had a fuel-efficient two-cylinder engine mounted at the rear, as well as a range of complementary safety features; it looked simple, but was surprisingly stylish and well finished—and the skeptics began to go quiet.

Four years of design work were put into the task of developing the 1 lakh car, and the Nano is the result. So how did Tata go about creating such a low-cost car, and can the process be repeated elsewhere?

The engineering team zeroed in on three vital parameters. Everything about the automobile had to be focused on low cost, even if it meant a rethink of conventional design principles. Secondly, the Nano had to meet all statutory safety and emissions legislation, while also building in the provision to meet additional safety and other legislation issues that changing

requirements could throw at it. And, finally, the car had to have acceptable performance.

The Nano's price of Rs 1 lakh was the only certainty at the time the project was conceived. With the unusual advantage of starting from a completely clean sheet of paper, the car was designed and developed to keep manufacturing costs, material costs, and post-purchase running costs at the lowest possible levels.

Crucial in this was the design packaging, the term engineers use to define how all the mechanical components are arranged in relation to passenger and luggage space. The twin-cylinder engine's small size allows it to be positioned at the rear, reducing the complexity of pipes and services that run under the car; this in

turn allows the passenger compartment to extend further forward, giving more legroom for front occupants. As it is the rear wheels that are driven, the driveshafts and their joints can be simpler and lighter too. Cost savings are also made by eliminating the requirement for a hood grille as an air intake; instead, vents next to either side of the rear doors channel cooling air to the rear-mounted radiator.

The Nano's small width (just 59 inches) may make it somewhat tight on interior elbow-room, yet this was crucial to reducing costs, as the smaller car quite literally requires less raw material to build.

Another vital objective for engineers was to keep the weight down. Lower weight and a

Opposite
The Nano explores new ground in terms of simplicity, use of space and, of course, cost reduction. The twin-cylinder engine is under the trunk floor, simplifying the mechanical layout.

Above
The luxury version of the Nano has alloy wheels and other embellishments; it will form the basis for export derivatives of the car.

limited top speed mean that suspension systems and brakes can be smaller and lighter, thereby allowing lower consumption, a smaller fuel tank, and still less weight. Clever techniques used by Tata engineers to minimize weight include computer optimization of the body structure to provide a stiff body with strength in the high-load areas. The engineers did not over-engineer the structure, which, although it could have provided greater levels of refinement and performance, would have pushed the weight up again.

Most optimization goes on under the skin, but there are some examples of visible design details that point to structural optimization. The roof has longitudinal ribs that are not just a style feature but which also provide additional stiffness: this in turn allows engineers to use thinner sheet metal than they would otherwise have specified. The rear glass windshield is bonded to the tailgate, helping to maintain its structural rigidity while cutting down on the weight and also simplifying the stamping and blanking processes.

The battery is located under the driver's seat, helping to spread the weight optimally, while the radiator is placed at the rear on the right-hand side. Leading supplier Bosch played a major role in the development of the multipoint fueling system and the electronic management system that enables tight control on the fuel consumption of the two-cylinder 623 cc engine.

The actual management process of developing the Nano was also crucial to its success. Harnessing the collective experience of the designers, engineers, and those who will ultimately build the car helped to simplify the assembly process and was vital in reducing costs through the whole process. Time is money when it comes to building cars, so simple fixing methods and parts that self-locate onto others save time and speed up the production process.

The quality of the finished car was important if the Nano was to stand out from the competition and was not to be perceived as cheap by looking cheap. The exterior finish

quality is good, with consistent and tight gaps between the Nano's metal body panels, and precise integration of the bumper to the car body. Inside, a simple but well-finished one-piece plastic dashboard carries a central instrument and control stack. Other components that give a good perception of exterior quality include the clear-lens headlamps, the stalked outside rear-view mirrors, the tubeless tires, and the high-mounted stop-lamp.

Inside, plastic panels have been innovatively designed to eliminate the need for screws: they fit by just snapping on firmly. The instrument cluster displays only basic information but is attractively designed. How the car will stand up to prolonged use remains to be seen.

Small wheels and tubeless tires both help to trim weight, though the small wheels may adversely affect ride comfort over rough ground. The seats are simple yet comfortable, without any luxury but with an all-important low ticket price. The seat design was a clear battle to balance cost, comfort, and complexity.

Other cost-saving features include the lack of any exterior ornamentation, such as chrome or door rubbing strips: instead, molded-in lines set into the panels give some added character. Both bumpers are designed for ease of molding, their fixing being designed to help create tight panel gaps on assembly. There is a single large windshield wiper at the front, again reducing the number of parts required. The design integration of the headlamps—and in particular their

relationship with their surrounding panels—is such that the fit and finish appear very good despite the fact that the panels have to match all the way round to curved edges.

Perhaps the best accolade a new model can obtain is a compliment from a competitor. Renault–Nissan chief Carlos Ghosn has said that the Nano is the best example yet of such engineering capabilities and comes at a time when carmakers, and indeed most engineering

Opposite
Extreme simplicity distinguishes the Nano's interior, with all the controls and the single instrument cluster being housed in the central console; transmission is either four-speed manual or CVT automatic.

Above
A central objective of the Nano program was to keep weight and wind resistance down, allowing the use of a smaller engine; narrow tires and small wheels help on both counts.

companies, are straining every muscle to cut costs in product development. Renault itself is working with Indian motorcycle manufacturer Bajaj Auto to develop its own low-cost car pitched slightly higher in price. Other makers monitor the developments with great interest.

The Nano, a neatly designed four-seater, is a car that, given time, stands to transform the lives of many thousands of people in India and across Asia and other developing markets. Although the Tata group is based in India, it is internationally minded and has made many acquisitions outside of India—not least of which is the Jaguar Land Rover group, which it bought from Ford in the first half of 2008. Ratan Tata has let it be known that export markets will figure in his plans for the Nano once domestic sales have reached a satisfactory level; even Europe is not out of the question at some point. Indeed, as evidence of his intent, two examples of the Nano were

displayed at the Geneva show in 2008. Longer term, Tata expects to be building around a million Nanos a year, and localized assembly in other regions could boost that figure still further.

Whatever the final production numbers prove to be, the Nano has already caught the eye of many industry executives and is being hailed as a new benchmark in low-cost vehicle design. As such, it could take the new prime spot in a hypothetical people's car hall of fame, an honorable tradition of designing and building cars for the masses that began with the Ford Model T in 1908. Other cars in that distinguished collection would include the little Austin Seven of 1922 (the first British car under £100), the Volkswagen Beetle, the Citroën 2CV, and two Fiat 500s—the 1936 Topolino and the Nuova Cinquecento of 1957. Included, too, would be the postwar Morris Minor, the original Austin-Morris Mini, and perhaps even the Renault 4.

Opposite, clockwise from top
A century of people's cars: the Fiat Nuova 500 of 1957 was a spectacular success, its moderate price giving millions of people mobility they had never had access to before; the Morris Minor sold steadily for two decades though never achieved true people's-car status; the Volkswagen Beetle, which had its origins in the 1930s, remained a popular global icon throughout the 1950s and 1960s; and the Ford Model T put America on wheels thanks to the moving assembly line and the populist pricing its efficient production permitted.

Above
A tall, narrow profile giving good headroom for four adults is a clear feature of the Nano's design.

A – Z of New Models

Audi A4

Engine	3.2 V6 (1.8 in-line 4, and 2.0, 2.7, and 3.0 V6 diesel, also offered)
Power	198 kW (265 bhp)
Torque	330 Nm (243 lb ft) @ 3000–5000 rpm
Gearbox	6-speed manual
Installation	Front-engined/all-wheel drive
Front suspension	Multi-link
Rear suspension	Multi-link
Brakes front/rear	Discs/discs
Front tires	225/65R16
Rear tires	225/65R16
Length	4700 mm (185 in.)
Width	1826 mm (71.9 in.)
Height	1426 mm (56.1 in.)
Wheelbase	2808 mm (110.6 in.)
0–100 km/h (62 mph)	6.2 sec
Top speed	250 km/h (155 mph)
Fuel consumption	9.2 l/100 km (25.6 mpg)

It would be tempting to see the new Audi A4 as just a freshening-up of the old design, with a couple of upgraded details and a different shape to the rear lights. Yet it is very much more than that, for this is one of the biggest generation steps in the A4's history. In fact, it would be more accurate to think of it as a four-door version of the A5 coupé, launched in spring 2007.

That car sought to bring Audi's handling responses and steering up to the level of the acknowledged market leader, BMW; Audi knew this could be achieved only by moving the mass of the engine further back in the chassis—an operation that required an extensive redesign of the transmission. As a consequence, the Audi now has much less front overhang—just like a BMW. As with the A5 coupé, the A4 has a strong, powerful stance on the road, with wide-set wheels fitting closely in their arches, and the smoothly arched cabin contributing to a feeling of spaciousness and luxury.

The waistline rises, as it did on the previous model, but now falls away slightly to the rear to give a more dignified look; the side feature line, beginning at the rear corner of the headlight, arches gently rearward to pick up the top edge of the taillight, while a second, stronger crease in the lower door sweeps upward to form the top of the rear bumper. The front is distinguished by the familiar Audi grille and complex, technical headlamps with built-in LED signature lighting that provides a classy and distinctive identity for the A4 on the road. The rear is neat, elegant, and attractive, with a steeply sloped screen and subtly different lamp shapes from the A5. The high-class interior carries over its dashboard from the A5—again, no bad thing—and a new micrometallic oxide gray is added to the choice of inlay panel materials.

Audi Cross Cabriolet Quattro

Design	Claus Potthoff
Engine	3.0 V6 Tdi
Power	176 kW (236 bhp)
Torque	500 Nm (369 lb ft)
Gearbox	8-speed automatic
Installation	Front-engined/four-wheel drive
Front suspension	Multi-link
Rear suspension	Multi-link
Brakes front/rear	Discs/discs
Front tires	265/35R21
Rear tires	265/35R21
Length	4620 mm (181.9 in.)
Width	1910 mm (75.2 in.)
Height	1630 mm (64.2 in.)
Wheelbase	2810 mm (110.6 in.)
0–100 km/h (62 mph)	7.2 sec
Top speed	240 km/h (150 mph)
Fuel consumption	7.3 l/100 km (32.2 mpg)

Los Angeles showgoers had been expecting Audi to reveal its long-anticipated Q5 SUV, its rival to the BMW X3; instead, the rapidly accelerating German premium carmaker lifted the lid on a flashy convertible—aptly enough for the sunshine state.

But the Cross Cabriolet is a cabrio with a difference: it is an off-road open car. A curious blend of sports car, sport utility vehicle, and convertible, it takes the already questionable logic of the Shanghai show Cross Coupé and propels it into a niche that is narrower still—if indeed there is niche there at all.

Yet rather than simply adapt the Shanghai design for open-air duty, Audi allowed its California designers a different take on the concept: the Cabriolet is longer between the axles and between the bumpers than the coupé, and the design treatment is more extreme.

Big, hunky wheels, jacked-up suspension, and a tall grille with multiple vertical slats emphasize the car's height, and the body creases are much sharper and more pronounced than is typical for Audi. A narrow waistband runs straight from front to rear, the crease bulging out over the wheel arches (like the classic ur-quattro) to give a wider stance on the road. Large gray sills help visually to divide up the height of the car and carry an aluminum step plate that echoes a similar structure in the lower front bumper.

The high-set rear is simple in its graphics but more complex in its curvatures: the lights are lozenge-shaped and wider toward their outboard edges, while the shoulder crease forms a spoiler lip to finish off the rear deck—perhaps an appropriate term for a rear end that has a slightly boatlike look.

This type of vehicle has no close equivalent in current production, though such manufacturers as BMW and Land Rover are approaching 4x4 coupés from an SUV direction. But, as to whether there is demand for a sports convertible that can cross rocky terrain, that is another matter altogether.

Audi Cross Coupé

Engine	2.0 in-line 4 diesel
Power	152kW (204 bhp)
Torque	400 Nm (295 lb ft) @ 2000–3500 rpm
Installation	Front-engined/all-wheel drive
Front suspension	MacPherson strut
Rear suspension	Multi-link
Brakes front/rear	Discs/discs
Front tires	245/45R20
Rear tires	245/45R20
Length	4380 mm (172.4 in.)
Width	1820 mm (71.7 in.)
Height	1600 mm (63 in.)
Wheelbase	2600 mm (102.4 in.)
Fuel consumption	5.9 l/100 km (39.9 mpg)

The Cross Coupé quattro is one of a multitude of concept vehicles shown by Audi at international auto shows. Each of these vehicles explores new potential market niches for the brand: the role of the Cross Coupé is to map out a design direction for a small-to-medium SUV to compete with the Land Rover Freelander and the Toyota RAV4.

This is a vehicle built on the same platform as the Volkswagen Tiguan, itself derived from the Golf and closely related to the Audi A3. It is immediately obvious that Audi's take on this type of vehicle is a much more sporty one: the Cross has the higher stance of an SUV, but the fastback rear gives it a coupé profile and the bold grille is aggressive in its shaping and sharp metal detailing. The LED headlamps appear insignificant in comparison with the dominant grille. The sculpted bodywork highlights tough-looking sills, emphasizing the machine's off-road credentials, though the shape of the body itself is hardly radical.

Detailed exterior points, such as the rear lights, reference the existing large Audi SUV, the Q7. The concertina-style fold-back fabric sunroof is a nod to older cars from an era when driving was more carefree. The pale leather interior, too, harks back to earlier times, though the execution is extremely contemporary and appears production-feasible. Different types of leather, ranging from the saddle-brown seat facings to the pale dashboard and door panels, create an appealing contrast; apart from subtle chrome highlights around the instruments, little metal is visible. Unusually, ventilation is handled by a full-width perforated strip rather than by individual air ducts. Advanced technical touches include a system to project information—including text messages and even parking permits—onto different locations on the windshield.

Audi Metroproject Quattro

Engine	1.4 in-line 4; electric motor on rear axle
Torque	440 Nm (324 lb ft)
Installation	Front-engined/front drive; electric rear drive
Front suspension	MacPherson strut
Rear suspension	Multi-link
Brakes front/rear	Discs/discs
Front tires	225/35R18
Rear tires	225/35R18
Length	3910 mm (153.9 in.)
Width	1750 mm (66.3 in.)
Height	1400 mm (55.1 in.)
Wheelbase	2460 mm (96.9 in.)
0–100 km/h (62 mph)	7.8 sec
Top speed	201 km/h (125 mph)
Fuel consumption	4.9 l/100 km (48 mpg)
CO_2 emissions	112 g/km

There is nothing experimental or exploratory about this small Audi. In everything but name and the odd exotic detail, this is the forthcoming A1: Audi's bid for a slice of the premium small-car market presently monopolized by Mini.

Unlike the Mini, however, the Metroproject quattro is not retro in its design, unless you count the large, bold square grille with its checkered mesh, reminiscent of vintage racing Auto Unions. The rounded features of the compact two-door hatchback represent a logical distillation of Audi design code into a smaller format; only the grille is oversized. Yet there is a harder edge to some of the detailing than we are used to from Audi: the straight feature lines taken from the top corners of the grille form heavy eyebrows over the headlights, giving the car an angry, staring expression, and the lower edge of the clamshell hood establishes a strong horizontal line that runs rearward, through the aluminum fuel filler cap, and into the top edge of the rear lights, where it again results in a sulky, brooding look.

On the side, the dominant feature is the semicircular, bright aluminum arc that forms the A-pillar, cant rail, and C-pillar of the upper architecture: this recalls the graceful arch of a wide bridge and lightens the presence of the glasshouse. Overall, this is a well-proportioned and beautifully resolved small car with a distinctive character and, on the show concept at least, distinctive engineering.

A conventional 1.4-liter turbo and supercharged engine power the front wheels, while an electric motor—intriguingly visible through a Perspex panel in the trunk floor—allows hybrid operation and a range of about 60 miles on battery power alone. Buyers choosing the A1 will have a cockpit very similar to that of the exotic R8 supercar, the steering-wheel design in particular; and a neat touch is a dashboard air vent design that mimics the impellor motif of the chrome-plated wheels.

BMW 1 Series Coupé

Design	Chris Bangle
Engine	3.0 in-line 6 (2.0 in-line 4, and 2.0 diesel, also offered)
Power	226 kW (306 bhp) @ 5800 rpm
Torque	400 Nm (295 lb ft) @ 1300–5000 rpm
Gearbox	6-speed manual
Installation	Front-engined/rear-wheel drive
Front suspension	MacPherson strut
Rear suspension	Multi-link
Brakes front/rear	Discs/discs
Front tires	215/40R18
Rear tires	245/35R18
Length	4360 mm (171.7 in.)
Width	1748 mm (68.8 in.)
Height	1408 mm (55.4 in.)
Wheelbase	2660 mm (104.7 in.)
Track front/rear	1470/1497 mm (57.9/58.9 in.)
Curb weight	1560 kg (3439 lb)
0–100 km/h (62 mph)	5.3 sec
Top speed	250 km/h (155 mph)
Fuel consumption	9.2 l/100 km (25.6 mpg)
CO_2 emissions	220 g/km

Not so much a coupé as a compact two-door sedan, this variant of the 1 Series has two objectives: first, to rekindle the spirit of the classic 2002, the iconic compact sports sedan of the 1960s; and, secondly, to provide a simpler and perhaps lower-cost entry point into the mainstream BMW range.

The shift to coupé format further distorts the already odd proportions of the 1 Series. Though the design is identical to the hatchback from the front grille to the A-pillar and windshield, the much shorter cabin leaves a short, flat rear deck that serves to offset slightly the strongly cab-rearward look of the hatch; the rear overhang is actually longer than that of the hatch, but this still does not detract significantly from the 1's characteristic long-hooded look.

Links with the 2002 are suggested by the unfashionably upright rear screen, allowing a separate trunk lid rather than a hatchback tailgate, and by a strong side shoulder line that—at least from the door rearward—references the all-round feature line of the classic model.

The Coupé is able to achieve a clean and simple look, just like the 2002, through clever and quite complex surfacing, especially on the doors and the rear three-quarter panels, the latter curving in strongly to give a neat, compact tail. Frameless doors and the characteristic BMW kink in the C-pillar line give a clean, fresh DLO profile, though this car still shows much more of a direct link to the hatchback than the larger and sleeker 3 Series Coupé does to its sedan progenitor.

When the car is viewed from the rear, the strong inboard pull of the body sides makes for a powerful impression, but the rear-panel treatment is fussy and awkward, the trunk lid following the complex chicane-shaped profile of the rear lights' inner edges.

BMW CS

Two elements lay behind the huge stir caused by the CS when it was unveiled at the Shanghai auto show in 2007. The fact that BMW had chosen a Chinese stage to give the world its first glimpse of the make's new top luxury identity was significant in itself, but no less significant—and what shocked commentators—was the fact that this new identity relinquished the large limousine look for something much more sporty.

So sporty, in fact, that at first glance the CS looks like a large grand-touring coupé, not a chauffeur-style sedan. This is, of course, a visual trick: the car is very wide—6 feet, 6 inches—and the superstructure is kept low, with all the features forward of the A-pillar drawing the eye towards the broad, aggressively thrust-forward BMW double grille.

Wide-set brake-cooling air intakes further exaggerate the wide stance: huge wheels fill the wheel arches so completely that suspension travel appears non-existent. The sides themselves are sculptural, in the idiom of current production BMWs, but add a shoulder over the rear wheel arches to strengthen the coupé roofline that begins to arc downward through the rear-door glass. The back window is drawn out almost horizontally, while the shape of the taillights is echoed in the exhaust outlets below.

The interior is a zone of great contrasts, with a steering wheel trimmed in black and white leather, black coverings to the dashboard top that sit on pale leather lower surfaces, and a nostalgic return to round ceramic control knobs within a bronze metallic dashboard. The general effect is of elaborate, classical precision.

While the CS, which revives a famous BMW designation, lacks the simple elegance of its 1970s namesake—and for that matter of the Aston Martin Rapide or the Maserati Quattroporte—it is certainly a clear signal that BMW means business in the four-door luxury coupé segment, which already includes Mercedes-Benz and will soon draw in Porsche too.

Design	Adrian van Hooydonk
Engine	5.0 V10
Installation	Front-engined/rear-wheel drive
Brakes front/rear	Discs/discs
Front tires	21 in.
Rear tires	21 in.
Length	5106 mm (201 in.)
Width	1978 mm (77.9 in.)
Height	1367 mm (53.8 in.)
Wheelbase	3142 mm (123.7 in.)
Track front/rear	1596/1682 mm (62.8/66.2 in.)

BMW X6

Design	Chris Bangle
Installation	Front-engined/all-wheel drive
Brakes front/rear	Discs/discs
Front tires	21 in.
Rear tires	21 in.
Length	4877 mm (192 in.)
Width	1979 mm (77.9 in.)
Height	1696 mm (66.8 in.)
Wheelbase	2933 mm (115.5 in.)
Track front/rear	1665/1665 mm (65.6/65.6 in.)

Designs that marry the seemingly contradictory notions of coupé sleekness with the tough, go-anywhere bullishness of a 4x4 have appeared as concept studies at motor shows for several years now. But it has taken the highly respected and certainly not fashion-led name of BMW to commit to manufacturing just such a machine.

Built on the established engineering platform of the X5 off-roader, the X6 is by turns absurd, intriguing, and perplexing. It sits high off the ground on huge wheels, yet the strong wedge shape to the body gives it a racy, tipped-forward look.

The picture is further confused by the peculiar profile of the coupé-like cabin area. Rising from the fast A-pillar to a high point above the driver's head, the DLO then drops sharply over the rear seating area to form a triangle with its acute angle at the rear. Placed on top of a waistline rising steeply to the rear, this makes the roofline roughly horizontal, in effect neutralizing the wedge of the lower body.

The rear window, with its sharp upper edge, is coupé-like, too, yet it sits above a very high-set tail, accentuating the plunge of the wedge when looking forward from the rear quarters. The chrome band through the rear lights is a classy touch. Against this, the front-end design is relatively conventional BMW.

Having created what it described at the time as the world's first sports activity vehicle, in the shape of the X5, BMW showed that it could make an effective off-roader a good on-road handler, too. In the X6 BMW has sought to combine qualities that no one would normally think of mixing; the result is a design that shakes up not only accepted notions of proportion, but also those of purpose and practicality. Yet BMW insists it has discovered a new market niche, and it has a habit of being right.

Buick Riviera

Design	Ed Welburn
Brakes front/rear	Discs/discs
Front tires	21 in.
Rear tires	21 in.
Length	4710 mm (185.4 in.)
Width	1940 mm (76.4 in.)
Height	1415 mm (55.7 in.)
Wheelbase	2870 mm (113 in.)
Track front/rear	1645/1635 mm (64.8/64.4 in.)

The Riviera is yet another concept that underlines the growing importance of China to Western carmakers. Buick was the brand that spearheaded GM's initial push into China; now the fact that the American giant entrusted the design of a key model to its Chinese design center is proof of the maturity of its whole China operation.

For Buick, this is something of an emotional return to a brand-defining nameplate that has been absent from the lineup for eight years. Rivieras have always been four-seater, two-door coupés, and this concept remains faithful to the template, apart from the very dramatic difference that the two doors are large gullwing affairs that give access to front and rear seats.

Under the concept's handsome skin sits a hybrid powertrain that will enter volume production in China in 2008; yet most eyes will be drawn to the tightly stretched shape, which plays on a contrasting mix of smooth, elegant curves and sharper details, often picked out from the pale blue paintwork in chrome. One such detail is a big, smiling version of Buick's trademark waterfall grille, flanked by sharp-edged headlights that flow back into the horizontal surface of the hood.

Particularly novel at night are the doors, which, when fully open, shine the word "Riviera" onto the ground. Indeed, the design makes ample use of lighting effects, including ambient light strips and icy-green backlighting under the doors.

Inside, the Riviera serves up more familiar concept-car fare in the shape of a futuristic look with sleek and streamlined forms, plentiful chrome and pale leather on the upholstery, and heavy use of aluminum in the center of the dashboard. Cream and blue hues represent earth and water, according to Buick.

China is Buick's largest market worldwide, and the Riviera is a sure sign that the center of gravity of the brand has moved from Michigan to Shanghai.

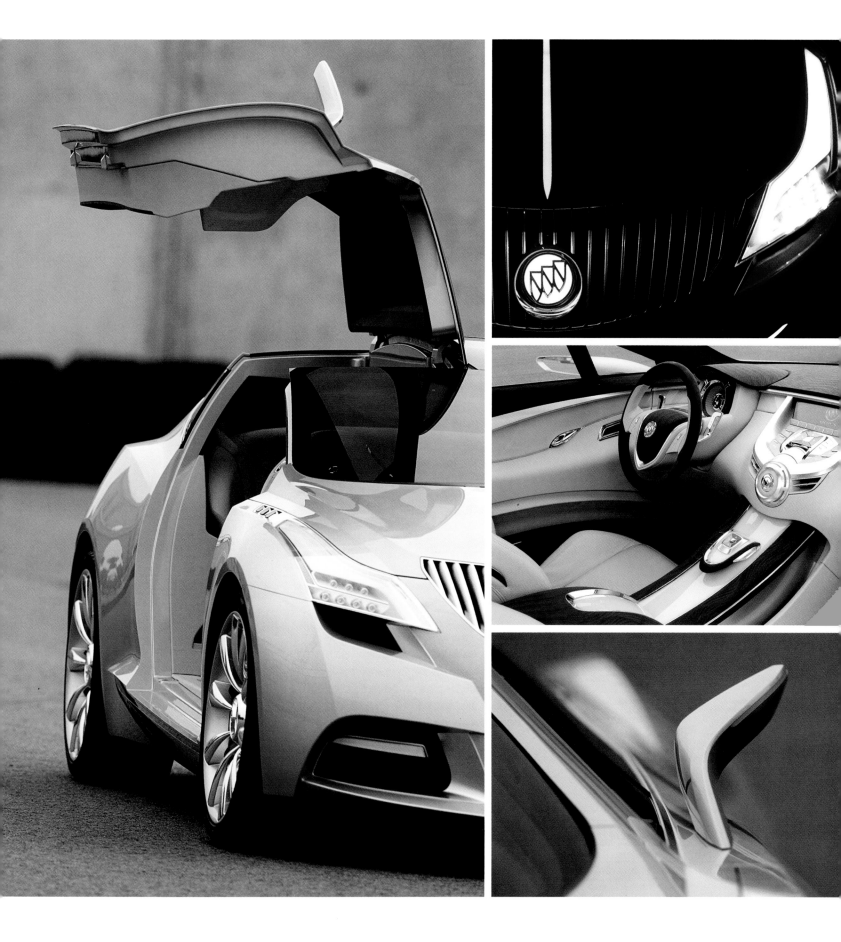

Cadillac CTS Coupé

Design	Clay Dean
Engine	3.6 V6 (2.9 diesel also proposed)
Power	227 kW (304 bhp)
Gearbox	6-speed manual
Installation	Front-engined/rear-wheel drive
Front suspension	MacPherson strut
Rear suspension	Multi-link
Brakes front/rear	Discs/discs
Front tires	20 in.
Rear tires	21 in.
Length	4750 mm (187 in.)
Width	1870 mm (73.6 in.)
Height	1390 mm (54 in.)
Wheelbase	2880 mm (113.4 in.)
Track front/rear	1570/1575 mm (61.8/62 in.)

The Coupé version of Cadillac's midsize CTS was one of the few cars to come as a genuine surprise to the crowds attending the 2008 Detroit show. The surprise was a very pleasant one: the rakish, aggressive exterior of the design was widely praised, and the bosses of parent company GM must have returned to their nearby headquarters full of determination to put the Coupé variant into production.

Unusually for a coupé derivative, the CTS retains the full wheelbase of the donor sedan, though it is lower and, thanks to the reduced rear overhang, shorter. The design retains the tight radii and the accurate, machined look of the sedan but comes across as a more emotional treatment. A complex set of interrelationships between lines and surfaces engenders a powerful technical feel; particularly characteristic are the sharp points and angles—especially the acute triangle formed by the rear quarter-window. This, and the sharp cut made by the forward edge of the side window, jar the eye in a provocative manner.

The rear end is high-set, with an almost horizontal rear window. The vertically stacked rear lights are angled inward, highlighting the shape of the trunk lid, and then fold around the sharp edge of the rear deck to lie parallel with the rear glass.

While the exterior is innovative in its form and proportion, the interior is more conventionally designed, being a simple transplantation of the sedan's themes. However, materials, textures, and detailing have been enhanced, and with perforated yellow suede seat facings and yellow stitching on the dashboard and steering wheel there is a very different ambience. A mixture of gloss black panels and matt leather makes for a contemporary contrast with the buttons in gray and the satin chrome dial surrounds. Retaining the sedan's wheelbase ensures, unusually for this class of vehicle, that rear-seat accommodation is more generous than the normally cramped 2+2 offering.

Cadillac Provoq

Design	Ed Welburn
Engine	Hydrogen fuel cell and lithium-ion battery
Power	148 kW (199 bhp)
Gearbox	None
Installation	All-wheel drive through central front electric motor and rear hub motors
Front suspension	MacPherson strut
Rear suspension	Multi-link
Front tires	205/60R21
Rear tires	205/60R21
Length	4580 mm (180.3 in.)
Width	1850 mm (72.8 in.)
Height	1703 mm (67 in.)
Wheelbase	2906 mm (114.4 in.)
Track front/rear	1639/1639 mm (64.5/64.5 in.)
0–100 km/h (62 mph)	8.5 sec

Though the Provoq grabbed the headlines on the strength of its hydrogen-fed fuel-cell powertrain (indeed, it was first presented not at Detroit but at the Las Vegas Consumer Electronics Show shortly beforehand), it has a deeper significance for Cadillac's mainstream model policy.

For this is the shape of Cadillac's upcoming medium crossover, fruit of parent company GM's vision of an SUV that is smaller and more environmentally responsible but which still distills Cadillac's premium style and values. With a conventional powertrain, the production Provoq will do battle with the BMW X3, the Land Rover Freelander/LR2, and the Mercedes GLK, the latter also making its debut at Detroit.

The Provoq shares its architectural underpinnings with the new Saab 9-4X, but the contrast between their exterior body styles could hardly be greater. The Saab is comparatively smooth and gentle, while the Cadillac is all sharp angles, straight lines, points, and edges. The treatment is bold and angular, with Cadillac's contemporary sharp creases giving a strong dynamic feature line plunging down as it runs forward from the rear lights, through the door handles, and to the air vents just behind the wheel arches; the line continues forward to form the sharp shelf of the front bumper as it wraps around the front fender.

Visually, there is a lot happening at the front, with Cadillac's signature boxy grille being thrust forward in a V-shape; louvres within the grille close at higher speeds in order to improve aerodynamic efficiency, opening again to cool the fuel-cell system as the vehicle slows down.

The rear is dominated by Cadillac's characteristic vertical rear lights and an elegant rear window hooded under a neat integrated roof spoiler. Inside, the forms deliberately correlate with those on the exterior, and the blue line running round the cabin is a subtle reference to the electricity flowing through the vehicle's fuel-cell powertrain.

Chevrolet Beat

Beat, Groove, and Trax are three GM design concepts for a young, urban car that were first aired at the 2007 New York motor show; they are sister designs based on a future small-car architecture and were conceived at GM's Asia Pacific design center in Korea, where their likely GM Daewoo manufacturing plant is located. GM's idea in showing three related but contrasting designs was to see which would best capture the public imagination; this in turn would guide the corporation in its decision as to which version to put into volume production.

Seven months later, at the Los Angeles show, GM announced the results of an online poll asking web-surfing gasheads to vote for their favorite among the three. "The Beat resonated with customers all around the world," said Ed Peper, Chevrolet's general manager. "Chevrolet was overwhelmed by the positive reaction to each of the 'triplets,' but the Beat was the clear winner."

Designed to be simple and free and easy, and to have a strong heart—even if it is only a 1.2-liter engine—the Beat has powerful graphical forms that combine to make it a visually stimulating shape; when viewed from the front the car is planted clearly on the road, with the wheels pushed wide out.

Diamond-shaped headlights accentuated in length are mirrored by wraparound rear lights, while the subtle creases above the front and rear wheel arches give a sense of movement, and such add-on details as lamps and spoilers give an opportunity for the personalization that is so important on a sporty young car.

The lurid green of the exterior is continued inside the vehicle, giving a wavelike movement to the dashboard to echo the exterior treatment. In contrast, smart black marks out the control and information-giving zones of the dash.

Design	Injo Kim
Engine	1.2 gasoline
Gearbox	Automatic
Installation	Front-engined/front-wheel drive
Front tires	17 in.
Rear tires	17 in.

Chevrolet Groove

In contrast to the Beat and the Trax, the Groove is the member of the GM city-car design trio that does not want to be seen as cute, cuddly, or small: hence a big, bullish look reminiscent of retro-style hot rods, complete with a prominent raised hood and sinister semi-matt black stealth paintwork.

"We wanted to get out of the mini-vehicles stereotype that is characterized by a weak, insecure, and 'cute' appearance, so that consumers can feel safe just by looking at the vehicle," said chief designer Jawook Koo of GM's Asia Pacific design center.

Few would dispute the imposing, masculine looks of the Groove, despite its compact dimensions and, under the voluminous hood, an environmentally unaggressive 1-liter diesel engine. Bold, sculpted wheel arches contribute to a mini-PT Cruiser impression, and closer examination shows that the door cuts are angled forward and the side crease downward to create a sense of motion.

The flat roofline, long glasshouse, and upright rear hatch bring a spacious look, and exterior detailing shows much originality—in particular the intriguing small tubular pods attached at the base of the side windows. These turn out to have indicator lenses to the front and rearview cameras pointing backward.

Contrasting with the distinctly retro style of the design's main surfaces is the thoroughly contemporary treatment of the rear end. Here, a smooth darkened-glass vertical tailgate extends down to bumper level and is flanked by large, upright light clusters incorporating big circular raised nacelles housing the brake and indicator lights. Beneath the license plate, recessed into the lower bumper, nestle twin chromed trapezoidal exhaust outlets—perhaps somewhat excessive for a mini city car powered by a 1-liter diesel engine.

Design	Jawook Koo
Engine	1.0 diesel
Gearbox	Automatic
Installation	Front-engined/front-wheel drive
Front tires	17 in.
Rear tires	17 in.

Chevrolet Traverse

This medium–large crossover SUV is Chevrolet's take on the platform already used by fellow General Motors brands Buick and Cadillac with the Enclave and Provoq respectively. As such, it relies on Chevy's new frontal template—the double-decker bowtie grille—to give it the necessary distinctiveness in relation to its stablemates.

The proportions are strong and the appearance bold, and as a long vehicle it is made more interesting and more dynamic by the kick-up in the waistline behind the C-pillars. A subtle side feature line runs through the door handles, droops at either end, and fades out before reaching the wheel arches; apart from this, however, there is little detail or focus on what is quite a plain, mathematical design.

The car's body lines are taut enough but, when seen from the side at least, fail to make the Traverse stand out from the crowd; the rear lamps offer something in the way of fresh organic curves and are reminiscent of Volkswagen design. Even so, the overall impression is bland and conservative from any angle except the front. At the rear, the shape of the tailgate glass mimics the kicked-up profile of the side windows, but the two do not visually flow together; instead, the substantial D-pillar creates a bulky impression around the rear quarter.

Inside, the same message of caution and conservative simplicity rules. The sculpted dashboard—in three shades of gray – flows smoothly into the doors and has the merit of being simple, uncluttered and not burdened with surplus chrome, the twin instrument bezels being the exception. The sloping center console blends neatly with the flow of the dashboard, while the three-row seating claims to offer enough room for adult space in every row.

The Traverse is undoubtedly the type of versatile family vehicle Chevrolet needs right now. Few are likely to object to it, but few are likely to remember it, either.

Design	Michael Burton
Engine	3.6 V6
Power	213 kW (286 bhp) @ 6400 rpm
Torque	346 Nm (255 lb ft) @ 5500 rpm
Gearbox	6-speed automatic
Installation	Front-engined/all-wheel drive
Front suspension	MacPherson strut
Rear suspension	Linked H-arm
Brakes front/rear	Discs/discs
Front tires	255/65R18
Rear tires	255/65R18
Length	5207 mm (205 in.)
Width	1999 mm (78.4 in.)
Height	1849 mm (72.8 in.)
Wheelbase	3020 mm (118.9 in.)
Track front/rear	1722/1712 mm (67.8/67.4 in.)
Curb weight	2234 kg (4925 lb)

Chevrolet Trax

As the urban off-roader of Chevrolet's micro-city-car design trio (the others being the Beat and the Groove), the Trax is perhaps the most unusual design, as well as potentially the most controversial.

The Trax comes across as tough, high-riding, and adventurous, despite its miniature proportions: it could have been conceived as a mini Land Rover. A two-tone color scheme deliberately differentiates between the contrasting functions of engineering and accommodation: the utilitarian wheels, chassis, and powertrain necessary for the car's mobility, and the passenger compartment that is directed at passenger comfort, convenience, and safety.

The exterior is a fundamentally geometric design built out of straight lines and gently curved surfaces; a good example is the curve of the hood sides as they flow rearward from the circular headlamps alongside the twin-deck mesh grille that has become Chevrolet's new frontal identity.

Pleasing details add to the authentic look: external catches recessed just below the A-pillars lock down the hood; a chunky spare wheel clings, Jeep-style, to the tailgate; and a businesslike aluminum roof-rack system signals that this vehicle is eager for hard, practical work and adventure.

Practical, too, are side door openings that are as large as feasible for such a small machine; the handles echo the simple style of the hood latches, though other details, such as the rearview cameras, appear flimsy by comparison. The bright metal skidplates mounted centrally on the front and rear aprons have an unauthentic look to them, and the color treatment—the brown front and rear ends appearing to have been dipped in milk chocolate—would be unlikely to play well with potential customers.

Well hidden from view is an innovative drivetrain, with the 1-liter engine driving the front wheels and an electric motor powering the rear, giving credible four-wheel drive as well as the potential for energy-saving hybrid operation.

Design	Sangyeon Cho
Engine	1.0 gasoline; electric motor
Gearbox	Automatic
Installation	Front-engined/front-wheel drive; electric drive to rear
Front tires	16 in.
Rear tires	16 in.

Chrysler ecoVoyager

Design	Greg Howell
Engine	Hydrogen fuel cell and lithium-ion battery
Power	200 kW (268 bhp)
Gearbox	None
Brakes front/rear	Regenerative braking
Length	4857 mm (191.2 in.)
Width	1915 mm (75.4 in.)
Height	1600 mm (63 in.)
Wheelbase	2946 mm (116 in.)
0–100 km/h (62 mph)	8 sec

The ecoVoyager is one of the first concepts from a major US automaker to demonstrate the real potential of fuel-cell power and electric drive in freeing up the fundamental architecture of the vehicle. As such, it gives an indication of what next-but-one generation family vehicles may be like in, say, 2020.

The exterior shape—basically a big, flowing bubble, tapering to a teardrop shape at the rear—gives little clue as to the engineering underneath, though the educated eye will be puzzled to note that the car's wheelbase is very long and there is almost no overhang at front or rear; given that passenger accommodation appears to take up almost the entire length of the vehicle, there seems to be no apparent space for the engine or transmission.

The answer, of course, is that the entire fuel-cell powertrain, the electric drive motor, the storage batteries, and the associated onboard electronics are all packaged below the cabin floor, leaving almost the complete length free for people and luggage. Indeed, the cabin is probably the best feature of the design, its four generous airline seats clad in calm white leather and its dashboard taking simplicity to unaccustomed extremes. The tops of the door panels sweep round to the base of the windshield to form the upper edge of the full-width electronic instrument strip; apart from column stalks at either side of the very simplified, almost 1950s-style steering wheel, there are no visible controls.

Externally, the ecoVoyager is less accomplished. It is sleek enough but the shape is unbalanced, with the visual mass pushed forward: the crisscrossing side feature lines do little to alleviate this; indeed, the rear haunch adds to the confused effect. At the rear, the beetle-backed ecoVoyager is again intriguing, though hardly beautiful, with its pointed tail, high-set rear window, and odd spine developing along the length of its roof.

Citroën C5

Design	Alexandre Malval
Engine	2.7 V6 diesel (1.8 and 2.0 in-line 4 and 3.0 V6, and 1.6, 2.0, and 2.2 in-line 4 diesel also offered)
Power	155 kW (208 bhp) @ 4000 rpm
Torque	325 Nm (240 lb ft) @ 1900 rpm
Gearbox	6-speed sequential automatic
Installation	Front-engined/front-wheel drive
Front suspension	Double wishbone
Rear suspension	Multi-link
Brakes front/rear	Discs/discs
Front tires	245/45R18
Rear tires	245/45R18
Length	4780 mm (184.3 in.)
Width	1860 mm (73.2 in.)
Height	1450 mm (57.1 in.)
Wheelbase	2820 mm (111 in.)
0–100 km/h (62 mph)	9.6 sec
Top speed	224 km/h (139 mph)
Fuel consumption	8.4 l/100 km (28 mpg)
CO_2 emissions	223 g/km

It is not for nothing that the publicity for the new Citroën C5 advertises it as the unmistakably German car that just happens to be made in France. Through its exterior design and interior architecture the C5 manages to convey a sense of solidity and quality that is typical of German cars, but that has consistently eluded French manufacturers. And in this so-called D segment, where volume manufacturers' models compete with products from such premium makers as BMW and Audi, the perception of quality is everything.

The C5 goes about this quality quest in several interesting ways. Its roofline is smoothly arched in the style of an Audi; its rear has something of the look of the VW Passat, though with important differences; and its body-side surfacing is made interesting by a concave undercut beneath the distinctive feature line, the section turning to convex further down in order to maximize the effects of light and shade. This is seen on the BMW 1 and 3 Series, for example.

Further promoting the classy, Germanic look are the solid-chrome pull-out door handles set into the feature line, and substantial chrome rubbing strips along the base of the doors and extending to the front and rear quarter panels. The use of brightwork elsewhere is sparing.

The front, with its long overhang and well-integrated double-chevron grille, is, however, pure Citroën, as is the concave rear window—borrowed from the larger C6—which enables the C5 to combine a three-box layout with a sleek look. There is a wagon, too, again with a well-resolved rear style.

Interiors are often a French weak point, but here again Citroën has put significant effort into giving the C5 a quality ambience. The discreet, soft-feel gray dashboard is restrained in its shaping and calm in its detailing, and the inclusion of Citroën's fixed-center steering wheel means that the majority of the minor controls are on the wheel, allowing the rest of the interior to be de-cluttered.

Citroën C5 Airscape

As suave and elegant as its stand-mate, the C-Cactus concept, was gawky and crude, the C5 Airscape is the template for a new generation of C5 sedans and wagons. This, a graceful four-seater coupé-cabriolet, will set the style for Citroën's bid for a slice of the action in the market for upper-medium sedans, where such premium products as the BMW 3 Series, the Audi A4, and the Volkswagen Passat currently hold sway.

Nowhere is the premium aura more vital than in the image-conscious world of coupés and convertibles. Fortunately, the Citroën—in common with its Renault Laguna convertible rival, also not yet confirmed for production—promises convincingly. There was rare unanimity among observers as to the pleasing balance of its profile, with the fast windshield echoed almost perfectly by the stretched-out rake of the rear screen; such details as the classy, long, tapering door handles scored well, too.

There is a discreetly sporty poise to the gently rounded front, the headlamps, the twin creases running back along the hood, and the wide-set wheels, but perhaps the most surprising feature is the highly successful rigid-panel folding roof. This is clad in carbon fiber designed to look like a traditional fabric roof, thus avoiding the slabby effect of too much body-colored metal and at the same time conveying the classy image of an Audi or Saab convertible.

The interior shows generous use of premium leathers and fabrics, with large pleated panels on the seat facings that seem to mimic the strap of a Rolex wristwatch. The three circular instruments initially appear conventional by Citroën standards but turn out to have annular needles that run round the dial rather than pivot from a central point.

The real test for Citroën will be to translate this quality feel into quantity production—and then to convince buyers that a Citroën can be just as good as an Audi.

Engine	2.7 V6 diesel
Power	155 kW (208 bhp) @ 4000 rpm
Torque	440 Nm (324 lb ft) @ 1900 rpm
Gearbox	6-speed automatic
Installation	Front-engined/front-wheel drive
Brakes front/rear	Discs/discs
Front tires	19 in.
Rear tires	19 in.

Citroën C-Cactus

Engine	1.6 diesel hybrid with 30 bhp electric motor
Power	52 kW (70 bhp)
Gearbox	5-speed automatic
Installation	Front-engined/front-wheel drive
Front tires	205/45R21
Rear tires	205/45R21
Length	4200 mm (165.4 in.)
Width	1800 mm (70.9 in.)
Height	1490 mm (58.7 in.)
Wheelbase	2800 mm (110.2 in.)
Curb weight	1109 kg (2445 lb)
Top speed	150 km/h (93 mph)
Fuel consumption	2.9 l/100 km (81.1 mpg)
CO_2 emissions	78 g/km

Glamour and good looks were not high on the list of priorities when Citroën drew up the design brief for this unusual concept car. Instead, the C-Cactus—so called on account of its ultralow consumption—is an entirely rational exercise aiming at reducing the cost and complexity of a medium-sized hatchback to the point where a super-economical diesel hybrid powertrain can be added and the sale price will be no higher than for a conventional car with a conventional engine.

Given that the cost of a diesel hybrid powertrain is several thousand euros more than that of a standard engine, Citroën designers had to employ radical thinking to take that much cost out of the basic vehicle structure. Accordingly, Citroën abandoned features that were not essential to comfort or safety, simplified major components—the doors consist of two parts rather than the usual twelve—and used the same component in multiple locations. The front body panel around the headlamps, for example, is repeated on the tailgate, and cutouts in the interior door stampings are used instead of add-on embellishments.

With function coming before form, the exterior looks purposeful rather than attractive, though the unpainted doors are certainly distinctive; there is no conventional hood opening, only a small access hatch; and the rear lights are translucent, allowing the driver to see through them for better rear vision.

It is in the interior that the most dramatic steps are seen. The dashboard has been replaced by simple controls within the steering wheel and an air-conditioner block in the center of the cabin, and the windows do not wind down but have small opening panels.

The C-Cactus may not be pretty, and many of its novel recycled-material ideas may not work in production practice, but it does just what a good concept car should do: propose ingenious solutions that set everyone thinking.

Dacia Sandero

Engine	1.6 in-line 4 (1.4, and 1.5 diesel, also offered)
Power	67 kW (90 bhp)
Gearbox	5-speed manual
Installation	Front-engined/front-wheel drive
Front suspension	MacPherson strut
Rear suspension	Torsion beam
Brakes front/rear	Discs/discs
Length	4020 mm (158.3 in.)
Width	1750 mm (68.9 in.)
Wheelbase	2590 mm (102 in.)

The Sandero is Renault low-cost subsidiary Dacia's take on a roomy family hatchback for sale in Eastern and Central Europe, South America, and Africa. Its mission is to partner the larger Logan, the remarkable success of which surprised the entire auto industry—barring, perhaps, parent company Renault—and which has forced major carmakers to rethink their strategies.

The Sandero shares many of the Logan's under-the-skin parts, helping to keep costs down, but is built (in Romania and, eventually, other locations) as a five-door hatchback—a body style favored mainly in Western Europe. The Logan is a four-door sedan or station wagon, styles that tend to be preferred in developing markets.

Unsurprisingly, there is little visual link between the two; the Sandero is crisper and slightly more modern, but it still looks dowdy and old-fashioned by the latest European and Japanese standards. This is evidenced in the flat sides and boot, the undersized wheels and, curiously, the deep side windows. The roof-mounted antenna is a further dated detail.

The grille and large headlights are well integrated in the modern idiom, and the narrow shelf that tracks the outer edge of the lights is a fresh detail. The rest, however, is simple, plain, and lacking distinction, though an odd detail is the half-moon depression stamped into each side's doors. This catches the light under certain conditions to link the front wheel arch with its counterpart at the rear.

The interior, though finished in plastics that to Western European eyes may look shiny and utilitarian, is based on previous Renault designs and is simple, practical, and inoffensive—a comment that could usefully be applied to the entire vehicle.

The point about the Sandero, of course, is that it is conceived around maximum value for minimum cost: it is a fresh design in a segment where ancient models have predominated for too long, and as such it represents a major step forward.

Daihatsu Mud Master

Here is a bold little concept that promises to do everything one would expect it to do, given that its name is Mud Master.

Effectively an evolution of the familiar "kei-class" micro-delivery vans that Japan has supplied to the rest of the world, the Mud Master is toughened, raised, and endowed with four-wheel drive to give it a go-anywhere ability, and its cargo bay is freshly rethought to allow both sides as well as the tailgate to open, giving unparalleled access to the payload within.

The Tokyo show concept was displayed with a brace of mountain bikes and an array of LED lights inside the gullwing side doors, giving intrepid bikers the ability to fix their machines in a well-illuminated and sheltered space even in the worst of conditions.

On a technical level Daihatsu has ensured excellent 14½-inch ground clearance by using reduction gears in the hubs, allowing the wheel center-lines to be lower than those of the axles. Front and, especially, rear overhangs are kept very short, and the front under-bumper area is tucked in very closely to allow the Mud Master to tackle steep slopes and pull out of tight gullies; a hefty skidplate protects against damage at the front and links to a projecting sill moulding that acts as a step right round the vehicle. At the rear, a similar plate doubles as a useful step for access to the cargo bed. With all rear doors open the taillights remain in position, attached to the C-pillars, and the front doors (which carry Oakley branding) hinge in a similar fashion around the headlights.

Though outfitted for use by extreme mountain-bikers, this intriguing and compact concept could have a variety of other uses both in leisure pursuits and by search and rescue services, its only real drawback being the high cargo bed forced by the mid-mounted engine.

Engine	0.66 liter
Installation	Mid-engined/four-wheel drive
Front tires	16 in.
Rear tires	16 in.

Dodge Challenger

Design	Jeff Gale
Engine	6.1 V8
Power	317 kW (425 bhp) @ 6200 rpm
Torque	310 Nm (228 lb ft) @ 4800 rpm
Gearbox	5-speed automatic
Installation	Front-engined/rear-wheel drive
Front suspension	Short and long arm
Rear suspension	Multi-link independent
Brakes front/rear	Discs/discs
Front tires	245/45R20
Rear tires	255/45R20
Length	5022 mm (197.7 in.)
Width	1923 mm (75.7 in.)
Height	1448 mm (57 in.)
Wheelbase	2946 mm (116 in.)
Track front/rear	1600/1603 mm (63/63.1 in.)
Curb weight	1878 kg (4140 lb)
0–100 km/h (62 mph)	5.4 sec
Fuel consumption	14.7 l/100 km (16 mpg)

Freshly cut loose from its German Daimler owners, the Chrysler group is returning to its historic brand heroes in order to re-emphasize the Americanness of its heritage. And nothing could be more American than the 1970s-style muscle car, specifically the Dodge Challenger and its on-street rivals of the era.

But there is more than just nostalgia and sentimentality behind Chrysler's decision to turn its much-lauded 2006 Challenger concept into a full production car. The one that started it all, the Ford Mustang, is already going great guns as a reborn pony car, and Chevrolet's twenty-first-century Camaro is about to hit the road, so Chrysler cannot afford to miss out on the action.

Muscle enthusiasts will be gladdened to see that the production Challenger remains true in almost every respect to the 2006 concept, even down to its characteristic deep Hemi orange paintwork. Given that the concept was itself a faithful replica of the 1970s original, the authenticity of today's example is as good as guaranteed.

As with the original, every line and every surface spells energy and impatience: the broad hood drawn back from the wide slot of a grille, with a pair of round headlamps set at either edge; the dual black hood stripes running back past twin air scoops to the sleek, shallow windshield; the big, wide-set wheels that give a powerful grip on the road. Yet, most of all, it is the shape of the roofline and the sharp kick-up of the waistline over the rear wheels that mark this as a Challenger.

The flat rear deck, too, is faithfully reproduced, with the full-width rear lights filling the inset rear panel and a small spoiler added to the upper edge. Inside, however, the Challenger steers clear of the futuristic detailing of the concept, instead favoring a sporty all-gray environment, with the two most important instruments—speedometer and rev counter—placed directly ahead of the driver.

Dodge Ram

Design	Ralph Gilles
Engine	5.7 V8 (4.7 and 3.7 V6 also offered)
Power	283 kW (380 bhp) @ 5600 rpm
Torque	548 Nm (404 lb ft) @ 4000 rpm
Gearbox	5-speed automatic
Installation	Front-engined/four-wheel drive
Front suspension	Upper and lower A-arms
Rear suspension	Multi-link
Brakes front/rear	Discs/discs
Front tires	245/70R17
Rear tires	245/70R17
Length	5308 mm (209 in.)
Width	2016 mm (79.4 in.)
Height	1887 mm (74.3 in.)
Wheelbase	3048 mm (120 in.)
Track front/rear	1730/1741 mm (68.1/68.5 in.)
0–100 km/h (62 mph)	6.1 sec
Fuel consumption	14.7 l/100 km (16 mpg)

Dodge parent company Chrysler, newly divorced from its nine-year relationship with Daimler of Germany, bills the new edition of the Ram truck as a "game-changer—a truck that breaks from the herd, literally." The Ram pickup is something of an American icon, as well as one of Chrysler's core profit-earning products. Yet, though the 2009 model did literally break with the herd at its Detroit launch— triumphantly arriving in town surrounded by a herd of 120 longhorn cattle and breaking free into the media gaze— there is little in the design to suggest that it has broken free from the shape or the simplistic engineering that are so familiar to so many Americans. In fact, the new model seems more Ram than ever, with an even more exaggerated version of the already imposing, forward-leaning, big-rig-style frontal grille, complete with crosshair motif. Even the separate front fenders, which serve to emphasize the sheer height of the massive raised hood, are retained.

What is visibly different is the much-touted crew-sized cab, with four doors and a big rear bench seat: this gives Dodge an entry into a market segment accounting for half of all pickup sales. Inside, there has been a move toward providing a more comfortable, quality atmosphere, with soft-touch—yet still rugged—materials, heated and ventilated seats, a heated steering wheel, and in-floor storage for up to ten beverages as well as ice.

Sensing a growing family-use market, Dodge has also provided sensible lockable, weatherproof luggage stowage down either side of the cargo bed—essential when all seats in the cab are occupied.

Boasting of "shockingly fast performance" from its 380-horsepower, 5.7-liter Hemi V8 engine, Chrysler's claim of game-changer status for the Ram lacks immediate credibility. Unless, of course, the company is really referring to the light-duty diesel and the innovative two-mode hybrid versions that are promised for the 2010 model year.

Dodge Zeo

Design	Bill Zheng
Engine	Electric motor
Power	200 kW (268 bhp)
Installation	Rear-wheel drive
Front tires	23 in.
Rear tires	23 in.
Length	4389 mm (172.8 in.)
Width	1742 mm (68.6 in.)
Height	1290 mm (50.8 in.)
Wheelbase	2792 mm (109.9 in.)
0–100 km/h (62 mph)	6 sec

The formula of four doors, four seats, and zero-emission battery power sounds like the recipe for a sedate neighborhood shopper rather than for a youth-oriented, thrill-seeking sports car. But what Dodge is trying to do with this Zeo concept—arguably the finest of the three presented at the Detroit show by the Chrysler group—is to prove that environmental responsibility need not be a barrier to style, excitement, or driving fun.

Electric power gives the designer distinct advantages in planning the packaging of a new vehicle: the powertrain elements are much more flexible in their positioning and their interrelationship, and control mechanisms are handy electric cables rather than cumbersome mechanical linkages. This freedom has allowed Dodge to package four people into a low-slung rear-wheel-drive sports car, accommodate vast 23-inch wheels, and still emerge with a pleasing, aerodynamic coupé shape.

The highly sculptural tangerine-colored bodywork wraps tightly round the wheels in the manner of a sports-racing car, the fender tops rising well above the beltline and intersecting the narrow rear window halfway through its height. This elongated beltline kicks up just before the rear wheel arch and forms the reverse angle C-pillar directly above the rear wheels; squeezed by the broad wheels, the cabin tapers noticeably in width toward the rear.

Pale gray sills and bumper areas reduce the visual mass of the body, especially at the sides, further emphasizing its lean look. The horizontal grille/light band has a similar effect and is notable for its blue-illuminated signature Dodge crosshair motif, symbolizing the flow of electricity.

The light-themed interior is restricted to white and pale gray, with black carpeting and blue accents to controls and displays. Perhaps deliberately, it has been given the virtue of clean, uncluttered, and futuristic design. But most of today's sports car drivers would argue that, compared with the exterior of the vehicle, the inside is disappointingly devoid of sporty feel.

Fiat 500

Design	Frank Stephenson
Engine	1.4 in-line 4 (1.2 and 1.3 diesel also offered)
Power	73.5 kW (99 bhp) @ 6000 rpm
Torque	131 Nm (97 lb ft) @ 4250 rpm
Gearbox	6-speed manual
Installation	Front-engined/front-wheel drive
Front suspension	MacPherson strut
Rear suspension	Torsion beam
Brakes front/rear	Discs/discs
Front tires	185/55R15
Rear tires	185/55R15
Length	3546 mm (139.6 in.)
Width	1627 mm (64 in.)
Height	1488 mm (58.6 in.)
Wheelbase	2300 mm (90.6 in.)
Track front/rear	1413/1407 mm (55.6/55.4 in.)
Curb weight	930 kg (2050 lb)
0–100 km/h (62 mph)	10.5 sec
Top speed	182 km/h (113 mph)
Fuel consumption	6.3 l/100 km (37.3 mpg)
CO_2 emissions	149 g/km

It is no exaggeration to say that this is one of the most eagerly anticipated new cars of recent times. With the huge success of the Mini having revealed great public affection for revivals of classic designs, Fiat knew it would be onto a winner with a born-again version of the cute and cuddly Cinquecento that put the Italian nation onto wheels in the 1960s and 1970s. Not only that: a rapturous reception was more or less guaranteed because of the many thousands of fans from around the world who had a hand in shaping the final design through an interactive website.

The inspiration of the original shape, created by the legendary Dante Giacosa in the 1950s, is there for all to see; today's car is clearly much bigger and more solid—the need for decent interior space and a safe structure sees to that—but it is no less cute, and the charm of the small, eyelike headlights and small clamshell hood will be irresistible to many; so too will the beautifully executed retro details, such as the Fiat emblem with its mouselike whiskers, the sloping rear end, and even the graphics of the 500 logo itself. Clever stylistic tricks have allowed a thoroughly modern design to be at the same time fresh and nostalgic, in many ways even more successfully so than the Mini, for which designer Frank Stephenson was also responsible.

The real delight, however, is in the 500's interior. This is a skillful blend of the retro and the contemporary, examples being the minimalist dashboard hosting sophisticated digital displays, and the contrast of painted surfaces with the classic white detailing of the control panels and steering wheel, which itself houses modern switches for operating iPods, phones, or other in-car devices.

Overall, this is a beautifully judged design with a deservedly high emotional pull. It successfully steers clear of pastiche and is sure to put a smile on the face of anyone who encounters it.

Fiat Fiorino Panorama

The Fiat Fiorino Panorama is just one representative of a new and extensive family of small utility and light commercial vehicles developed jointly by Fiat and PSA–Peugeot Citroën. Peugeot's equivalents are the Bipper van and the Partner MPV, while Citroën's are the Nemo and the Berlingo. All compete with the new-generation Kangoo from Renault and an increasing number of offerings from Ford, Volkswagen, and other volume carmakers.

All versions of the new Fiat/PSA design are similar in their chunky and rugged appearance, especially at the front, where the big projecting grille and bumper assembly sticks out forward from the headlamps to make for a truly distinctive look. The resulting flat shelf in front of the lights and hood could conceivably be useful in protecting the headlights and improving pedestrian safety, but it does not count as an aesthetic plus point.

Differentiation between the three makes is less crucial in light commercials than in the pure passenger-car market. In the new design it is confined, at the front at least, to the emblem on the hood and the band around the projecting grille.

More variety is to be found in the styles of the various body treatments aft of the B-pillars, with the sharply rising window line of the Fiat Fiorino passenger version making for a distinctive side view. On all versions, the recessed windows in the front doors give a sense of strength, and the simple, functional interior is clearly laid out, with a circular theme for the instruments, air vents, and storage spaces.

These vehicles will exist in a variety of versions and sizes, ranging from the small vans used by such tradespeople as couriers and electricians to the bigger, family-oriented derivatives that seat five passengers and are for many buyers a viable, roomier alternative to the smarter and perhaps unnecessarily luxurious compact MPVs, such as the Opel Meriva and the Renault Modus.

Design	Fiat Professional
Engine	1.3 in-line 4 diesel (1.4 gasoline also offered)
Power	56 kW (75 bhp)
Gearbox	5-speed manual
Installation	Front-engined/front-wheel drive
Front suspension	MacPherson strut
Rear suspension	Torsion beam
Brakes front/rear	Discs/drums
Length	3860 mm (152 in.)
Width	1710 mm (67.3 in.)
Height	1720 mm (67.7 in.)
Top speed	157 km/h (98 mph)
Fuel consumption	4.5 l/100 km (52.3 mpg)
CO_2 emissions	119 g/km

Fioravanti Hidra

Design	Matteo Fioravanti
Engine	Fuel cell
Length	4675 mm (184 in.)
Width	1880 mm (74 in.)
Height	1290 mm (50.8 in.)
Wheelbase	2900 mm (114.2 in.)

This smooth, gently styled concept from Italian design house Fioravanti is in fact a development of the Thalia proposal displayed by the company at the 2007 Geneva show. But while the Thalia had a peculiar configuration with a raised rear roof and a heavy, unbalanced double-decker look at the rear, the Hidra has a much more approachable and easily understandable design.

Unlike its predecessor, the Hidra gives off a clear sense of purpose: this is a fast and stylish four-seat grand tourer, with sports station-wagon overtones in its extended rear roofline and provision for a tailgate within the near-vertical wraparound rear window.

Restrained curves and a discreet low-set air intake characterize the front, while at the side two strong—and unusual—feature lines dominate. The waistline, formed at the base of the windshield, runs horizontally rearward before sweeping up dramatically to roof level toward the rear of the back door; a lower feature line, beginning at axle level at the trailing edge of the front wheel arch, rises through the doors to meet the top of the rear wheel arch, and then bends upward to run parallel to the first line. The whole graphic is emphasized by the two-tone color scheme, with a darker color below the feature line and a paler one above it. The only surprising observation is that the pale body color is not continued across the roof to bridge the tops of the C-pillars.

In this way the lower line forms the C-pillar and also serves to frame the large wraparound rear window. The base of that window is formed by a rearward extension of a third feature line, running back from the inner edges of the headlamps and the shutline of the hood.

As was the case with the Thalia, the Hidra was displayed as a solid model without any interior; this one, however, shows rather greater promise.

Fisker Karma

Design	Henrik Fisker
Engine	Electric motor plus gasoline range extender
Gearbox	None
Installation	Front-engined/rear-wheel drive
Brakes front/rear	Regenerative braking
Front tires	22 in.
Rear tires	22 in.
Length	4970 mm (195.7 in.)
Width	1984 mm (74.2 in.)
Height	1310 mm (51.6 in.)
Track front/rear	1689/1720 mm (66.5/67.7 in.)
0–100 km/h (62 mph)	6 sec
Top speed	200 km/h (125 mph)

Extravagant designer-luxury limo meets high-profile Hollywood star with a green conscience: that is perhaps the best description of the Fisker Karma. Crafted by designer Henrik Fisker, who set up his own California coachbuilding company after he left Aston Martin, the Karma departs from Fisker's previous strategy of reshaping existing sports car designs from BMW and Mercedes-Benz.

Instead, the Karma is a home-grown four-door sedan owing nothing to any other carmaker; most importantly, and perhaps with an eye to the runaway success of the $100,000 Tesla two-seater sports car with the California celebrity set, it is a near-zero-emission plug-in hybrid. Scheduled, with a touch of optimism, to go on sale in late 2009, the Karma stands a good chance of becoming the world's first fully environmental luxury car, and thus the choice of wealthy, image-aware, green-leaning executives and movie stars from Silicon Valley to Sacramento.

The design certainly has the dramatic looks to fulfill that demanding role. First impressions are striking, with a long, low hood, a low beltline, a very low, light roofline, and slender pillars, finished off by a neat, short trunk. The look is more coupé than limousine, accentuated by the way the flowing bodywork wraps tightly round the large wheels. The balance is somewhat upset by the broad full-width grin of the grille—surely unnecessary for a vehicle that makes a virtue of being electrically powered. The rear, too, suffers from the conflict of too many differently shaped elements.

Nevertheless, Fisker has demonstrated that by the clever packaging of the electric elements—the lithium-ion batteries are racked in the center tunnel of the aluminum chassis—generous space for four can be found within the profile of a low-slung four-door coupé. There is even room for a small four-cylinder gasoline engine under the hood to allow the driver to top off the batteries should he or she stray beyond the 50-mile range available on each overnight plug-in of the batteries.

Ford Explorer America

Design	Freeman Thomas
Engine	3.5 V6 (2.0-liter four-cylinder also proposed)
Power	254 kW (340 bhp)
Gearbox	6-speed automatic
Front suspension	MacPherson strut
Rear suspension	Multi-link
Length	4851 mm (191 in.)
Width	2022 mm (79.6 in.)
Height	1768 mm (69.6 in.)
Wheelbase	2939 mm (115.7 in.)
Track front/rear	1737/1773 mm (68.4/69.8 in.)

Considering that it is the blueprint for the next generation of one of North America's top-selling and most conservatively engineered SUVs, the Ford Explorer America concept is little short of revolutionary.

In a single move, Ford is proposing the abandonment of most of the most sacred tenets of traditional SUV faith. It moves from the old-fashioned body-on-frame construction to a lighter, stronger, and more refined unitary structure; it dumps the gas-guzzling V8 in favor of a more frugal V6 as the top engine, with the most popular unit expected to be a turbocharged four-cylinder 2-liter; and it adopts a fuel-efficient six-speed transmission. Underneath, modern independent suspension replaces the crude live axles, while electric power steering takes the place of hydraulic.

Yet, unquestionably, the Explorer is still an SUV, even if it has become softer, rounder, and friendlier in form. Thanks to the chunky cant rails and generous wheel arches it is still strong-looking, though no longer aggressive; its nearly 16-foot length is disguised by the big, body-colored C-pillar, which sweeps round to end visually the passenger compartment, leaving the cargo area, with its higher beltline, to appear as a separate entity. The rear, too, is curved and smooth rather than chunky, and the moon roof, extending over all six occupants, is a stylish touch. Unusual is the sliding rear door, offering easy access to the rearmost seats.

Front and rear display details that may not be carried over into an eventual production car. The signature three-bar Ford grille is now integrated into the bodywork, to the extent that it incorporates extensions that appear to cover the headlamps, while at the rear this not entirely successful treatment is echoed in taillamps consisting of a triangle of LEDs.

The interior is as gaudy as we have come to expect from some US concept vehicles; nevertheless, the significance of this revolution in America's most traditional backyard is not to be underestimated.

Ford F-150

The redesign of the F-150, America's favorite pickup truck for a generation or more, makes an interesting contrast with Ford's pretty radical proposed reshaping of the Explorer, the company's equally iconic SUV.

As one of North America's bestselling vehicles, the F-150 has always perceptively felt the pulse of the buying public. Ford has thus been reluctant to move the product on in any dramatic way, instead giving customers more of the same: more presence, more cab options, more engine and transmission choices, and, astonishingly, a different grille for each of the six different trim grades. The only parameter to have been reduced—albeit very slightly—is fuel thirst, which drops by 1 mpg.

However, given that the new F-150 was always going to be an evolutionary rather than a revolutionary development, it does represent a very intelligent and successful extension of the outgoing model's thinking. It is sharper and more contemporary, and is modern in the way it mixes plastic textures with body-color surfaces on the door mirrors and on the many different front-end treatments. The grilles themselves are skillful in the way they convey the various intended images of the vehicle, be it tough, sporty, or classy.

Many small details—such as the powered running boards and the built-in tailgate step—mark this as a well-thought-through, if unglamorous, design.

The square design theme continues inside, with thirty-plus stowage areas taking care of the requirements of a broad range of users craving everything from construction-site utility to city-center luxury. The blocky nature of the interior lends itself to a wide spectrum of different options to satisfy the needs of this exceptionally diverse clientele. Capitalizing on what Ford sees as a "how high do you want to go?" attitude among top-end pickup buyers, it has also added a Platinum trim grade, with everything from what Ford excitedly describes as "satin gloss lacrosse ash wood grain inserts" to embroidered logos on the seat backs.

Design	George Bucher
Engine	5.4 V8 (4.6 also offered)
Gearbox	4- or 6-speed automatic
Installation	Front-engined/four-wheel drive
Brakes front/rear	Discs/discs

Ford Fiesta

Design	Stefan Lamm
Engine	1.6 in-line 4 (1.2 and 1.4, and 1.4 and 1.6 diesel, also offered)
Gearbox	5-speed manual
Installation	Front-engined/front-wheel drive
Front suspension	MacPherson strut
Rear suspension	Torsion beam
Brakes front/rear	Discs/discs

Heavily trailed by the series of three Verve small-car concepts released at Frankfurt, Guangzhou, and Detroit and reviewed in this volume, the new Fiesta represents as big a step for Ford small-car design as the original Focus did in the medium segment back in 1998. Yet, perhaps because of the size of this step over the outgoing, very conservative Fiesta design, Ford has chosen to retain the Fiesta name and its associated positive recognition in the market.

The Fiesta remains faithful to the proportions and much of the detailing of the Verve hatchback concept, echoing its strongly forward-leaning wedge look, its high tail, and its hungry front. Nevertheless, some of the details have been softened: the nose, for example, is made less aggressive by the deletion of the chrome surround and mesh on the large inverted trapezoidal grille, and by the placement of the number plate in the center to break up its visual mass.

Kinetic design language gives the car a series of pointed graphical shapes to make it stand out from its softer-edged competitors: the long, spiky headlamps have points that extend a long way back, and the coupé-like impression is strengthened by the dynamic upsweep of the window line, along with the high-set taillamps visible from the side. Even the five-door version succeeds in projecting this impression. The rear aspect is no less aggressive than that of the Verve, retaining the neat trapezoidal lower rear lamps as well as the diffuser-style inset lower panel.

While the lurid lilac color highlights of the Verve's interior are passed over in favor of a cooler set of accents in green, the same provocative dashboard architecture is retained, along with the mobile phone keypad-style set of minor buttons in the center stack, set below the small recessed display screen.

Overall, Ford's Fiesta at last has the right ingredients to appeal to the highly fashion-conscious female buyer—just as the Peugeot 206 and 207 have done for so long.

Ford Flex

Design	J. Mays
Engine	3.5 V6
Power	194 kW (260 bhp)
Torque	333 Nm (245 lb ft)
Gearbox	6-speed automatic
Installation	Front-engined/front-wheel drive
Front suspension	MacPherson strut
Rear suspension	Independent coil over shock
Brakes front/rear	Discs/discs
Front tires	18 in.
Rear tires	18 in.
Length	5138 mm (202.3 in.)
Width	2030 mm (79.9 in.)
Height	1717 mm (67.6 in.)
Wheelbase	2995 mm (117.9 in.)
Track front/rear	1659/1661 mm (65.3/65.4 in.)

The Ford Flex is something of an unusual phenomenon in the auto industry: it is, in effect, a concept car that has been put into production with little, if any, perceptible change to its external appearance. The concept car in question was the Ford Fairlane, presented at the 2005 Detroit show, which we described as "sophisticated and dignified" in *The Car Design Yearbook 4*.

That air of sophistication comes from the Flex's simple, planar style, with a strong emphasis on the horizontal, an effect heightened by the clever use of dark-tinted windows and blackened pillars to hide the verticals and make the contrast-colored roof appear to float.

Ford describes the Flex as a crossover, aligning it with the current North American trend that has seen customers deserting SUVs in favor of this fast-growing segment. However, being long and low and seating seven in a three-row arrangement, the Flex has little obvious SUV in its genes; it presents itself as more of a cross between a minivan and a station wagon than as a 4x4 trying to be civilized.

Ford's designers have been successful in giving the Flex a classy, upmarket look, somewhat like a stretched and widened Range Rover; the only comparative letdowns are the brash bright-metal appliqué panel spanning the tailgate (again emphasizing the horizontal) and the ribbed effect stamped into the side doors.

Much is made of the quality feel of the car's interior; this is often a weak point of American designs, but the Flex successfully avoids the over-ornate look of many of its stablemates, even if it does not yet come up to European standards. Claimed to be an exclusive in the class is a refrigerator integrated between the two captain's-chair seats in row two: this, says Ford, can cool drinks 40 percent faster than a home fridge.

Ford Kuga

Design	Martin Smith
Engine	2.0 in-line 4 diesel
Power	100 kW (135 bhp)
Torque	320 Nm (236 lb ft) @ 2000 rpm
Gearbox	6-speed manual
Installation	Front-engined/front-wheel drive
Brakes front/rear	Discs/discs
Front tires	235/55R17
Rear tires	235/55R17
Length	4443 mm (175 in.)
Width	1842 mm (72.5 in.)
Height	1710 mm (67.3 in.)
Wheelbase	2690 mm (105.9 in.)
Curb weight	1613 kg (3556 lb)
Fuel consumption	6.4 l/100 km (36.8 mpg)
CO_2 emissions	169 g/km

Displayed in production form barely a year after the Iosis X concept that previewed its style, the Kuga is Ford of Europe's first internally developed SUV crossover; previous models, far more narrowly focused on pure off-road ability, were license-built designs from other Japanese companies.

The difference shows. The Kuga is very clearly a Ford, with its treatment of exterior surfaces and details—especially at the rear—strongly reminiscent of the successful new style established by the Galaxy, S-Max, and Mondeo. Compared with the Iosis X concept, however, the production Kuga is much less aggressive: gone are the angry scowl and menacing slit eyes of the concept's front, replaced by a more family-friendly look, while the steroidal, pumped-up wedge stance of the Iosis X has been moderated in favor of a sensible profile with a taller glasshouse and only slightly raised ground clearance.

Even so, the Kuga is well endowed with external design details intended to signal off-road toughness, despite the fact that the majority of examples sold are likely to be front-drive only. The front is punctured by no fewer than six air inlet apertures, with the large central intake featuring a prominent bright-metal stone guard along its lower edge. The design's sides show a strong rising feature line separating the concave lower and convex upper body surfaces, while at the rear a large skidplate appears to combine the roles of an aerodynamic diffuser and a protector for the bumper in extreme off-road maneuvers.

The five-sided rear lights neatly echo the style of the Galaxy and S-Max, but in contrast to those bigger, more passenger-oriented vehicles, the Kuga has a versatile split tailgate to facilitate loading of sports equipment.

The interior is very cautious compared with the screaming orange and black of the Iosis X: most of the elements are recognizable from other Ford models, and only some subtle orange piping provides a link with the concept.

Ford Verve

The aesthetic revolution that has swept through Ford's European model range, yielding such strong and successful designs as the S-Max and the Mondeo, has yet to reach the smallest vehicles in the company's lineup; as a consequence, the Fiesta and the once-trendsetting Ka (now a decade old) look painfully outdated.

The Verve, therefore, is a foretaste of the style and proportion of autumn 2008's Fiesta replacement, even if the actual production car is likely to be less exaggerated in its detailing. Even so, the production car will still be striking, echoing the Verve's bold inverted trapezoidal grille and its long, slender headlamps drawn back almost as far as the A-pillars. These features combine to make a strong visual statement, similar to current Peugeots.

The kick-up of the waistline toward the rear adds to the dynamic look and feel, while the triangular C-pillars have overtones of the Kuga, and Ford's designers have done a skillful job in resolving the complexities of the large number of surfaces coming together in this area. The sporty feel of the tail is enhanced by the blacked-out central diffuser panel and the trapezoidal reflectors set into the corners.

Inside, the busy center stack with its myriad angled silver keys shows mobile phones as a clear design influence; thankfully, this hectic theme is not continued through the rest of the cabin, which is already visually strident thanks to its lilac-and-gray coloring and pink pinstriping. The latter ties in with the bright lipstick hue of the exterior panels.

As with the Volkswagen Up! and last year's new Mini, the hatchback Verve shown at Frankfurt (pictured here) in 2007 was the first of a series of Ford small-car proposals. Two months later it had been joined by a sedan version, shown at Auto Guangzhou in China, with a third iteration—also a sedan, but with a fastback style—displayed at Detroit in January 2008.

Design	Chris Hamilton and Niko Vidakovic
Engine	In-line 4
Installation	Front-engined/front-wheel drive
Brakes front/rear	Discs/discs
Front tires	18 in.
Rear tires	18 in.
Length	4249 mm (167.3 in.)
Width	1730 mm (68.1 in.)
Height	1379 mm (54.3 in.)
Wheelbase	2489 mm (98 in.)
Track front/rear	1473/1455 mm (58/57.3 in.)

GMC Denali XT

Design	Warrack Leach
Engine	4.9 V8 with hybrid electric drive
Power	243 kW (326 bhp)
Gearbox	Electrically variable
Installation	Front-engined/rear-wheel drive
Front suspension	Multi-link
Rear suspension	Four-link
Brakes front/rear	Discs/discs
Front tires	255/35R23
Rear tires	285/35R23
Length	5207 mm (205 in.)
Width	1938 mm (76.3 in.)
Height	1587 mm (62.5 in.)
Wheelbase	3134 mm (123.4 in.)
Track front/rear	1651/1651 mm (65/65 in.)

From its dominant thrust-forward grille and big flat hood to its huge, wide-set wheels, the GMC Denali XT spells power and purpose. The impression is reinforced by the flattened low-rider roof—bringing to mind hot-rod specials—and the racy, forward-leaning C-pillar and tail. Who would guess, then, that this is not a high-powered sports car but a pickup—and a hybrid-powered pickup at that?

For the Denali, despite its eco-minded Two Mode hybrid powertrain, is not a vehicle for the faint-hearted, and at 17 feet in length it hardly makes a good case for economy or restraint. Such are the contradictions of the US car business, and the Denali is ample evidence that Detroit is harnessing advanced technology and alternative fuels (the XT is ethanol-compatible) in order to be able to continue building the big bruisers with which they feel most comfortable and for which they are famous.

On a design level, the Denali presses all the right buttons for a suitably bold, masculine allure. The prominent, raised hood creates strong shoulders around the front wheel arches: lateral slats in the grille further emphasize the width of the vehicle, while the lowness of the chopped, flat roof again increases the visual width.

The C-pillars angle down into the load bay to draw the eye to the fat rear wheel arches housing 23-inch wheels, and the drop-down tailgate wraps round to the sides, again for a massive look.

Inside its double cab, the Denali is a purposeful mix of high-quality saddle leather on the seats and dashboard and gray leather on the seat edgings, set off by chrome and polished-aluminum detailing. Innovative instruments have floating red-illuminated numerals set against a black background, which, to European eyes, are uncomfortably brash.

As a design exercise in pickup style the Denali is certainly impressive; as a vehicle to promote hybrid technology it is less so.

Honda Accord/Acura TSX

Engine	2.4 in-line 4 (2.0, and 2.2 diesel, also offered)
Power	142 kW (190 bhp) @ 6800 rpm
Torque	223 Nm (164 lb ft.) @ 4500 rpm
Gearbox	6-speed manual
Installation	Front-engined/front-wheel drive
Front suspension	Double wishbone
Rear suspension	Multi-link
Brakes front/rear	Discs/discs
Front tires	205/55R16
Rear tires	205/55R16
Length	4665 mm (183.7 in.)
Width	1760 mm (69.3 in.)
Height	1445 mm (56.9 in.)
Wheelbase	2670 mm (105.1 in.)
Curb weight	1398 kg (3082 lb)
0–100 km/h (62 mph)	8 sec
Top speed	227 km/h (141 mph)
Fuel consumption	9 l/100 km (26.1 mpg)
CO$_2$ emissions	214 g/km

When it comes to large cars, Honda is much more conservative than with its smaller models. The evolutionary nature of the updated 2009 Accord—together with its near-identical North American clone, the Acura TSX—proves the point. At first glance it is hard to distinguish the new model from the outgoing edition, especially the sedan model; with the wagon, a switch in role from frumpy cargo-carrier to sporty lifestyle leisure tool has allowed the new model a much sleeker, more Audi-like look at the rear.

Overall, the update is marked by clearer, crisper lines, sharper detailing, and the introduction of the currently fashionable flared wheel arches, those at the front being flattened to produce a semicircular feature line running outside the arch. The headlights are longer and narrower than before, and the angles on the grille and bumper are sharper.

Running rearward from the front wheel arch is a sharpened feature line that takes in the door handles and kicks up just above the rear lights to form the tip of the trunk-lid spoiler; on the wagon this line meets the rather larger rear lights. The window profile of the wagon is more interesting, rising toward the rear and emphasizing the angled shoulder above the feature line; on the sedan the rear is weightier and the strongly sloped C-pillars enclose the rear compartment protectively. The look here is slightly BMW, slightly Lexus—perhaps reflecting Honda's premium-alternative ambitions for the model.

The interior is again a cautious evolution of previous Accord practice, and makes a stark contrast to the adventurous dashboard rethink that characterizes the smaller Civic model. The Accord's dash is calm, clear and sober, with quality-feel instruments and controls, plus a solution for the central navigation screen enclosure that is arguably more successful than that of BMW or Mercedes. An element of style is added by the sweeping metal-finish beam that spans the center of the dash and sweeps down to meet the center console, providing useful extra storage slots at either side of the tunnel as it goes.

Honda CR-Z

A receptive and highly enthusiastic welcome was more or less guaranteed for the CR-Z's Tokyo show debut, given the substantial global fan base that still exists for Honda's iconic baby coupé of the late 1980s, the CR-X. Like its forebear, the CR-Z encapsulates the very latest in Honda engine technology—in this case a hybrid system offering low CO_2 levels yet strong acceleration performance; unlike the earlier car, the CR-Z has an intriguing exterior design that not only is neat and effective but also introduces new themes to the shape of compact sports cars.

Three stand-out features ensure that the CR-Z looks like no other small coupé. A large, rectangular air intake is thrust forward from the headlights and front fenders in the manner of a sports-racing car, the grille carrying the Honda emblem being well recessed; at the side, the sharply rising beltline sees the side windows plunge downward toward the tops of the front wheel arches and the rear side windows taper upward to meet the roofline with a subtle upward kink. And the rear is most unusual of all, with the near-vertical glass taking in the taillights at either side and dropping to the very base of the rear apron to house the twin exhaust pipes.

But it is when the car is viewed from the rear three-quarters that the real skill behind the CR-Z's body surfacing becomes clear: complex compound curves draw together the roof, rear pillars, rear fender, and body sides with a fascinating interplay of light and shade.

The steeply raked windshield, roof, and near-horizontal hatch glass form a single surface, while inside the cabin blue acrylic highlights the key elements. The CR-Z shown at Tokyo was a near production example, with Honda confirming shortly afterward that the model would launch as part of a global hybrid initiative in 2009.

If this is the shape of next-generation "green" cars from Honda, then even ultrachoosy sports-car drivers and those seeking maximum style have nothing to fear.

Honda FCX Clarity

Honda's FCX Clarity was the most-photographed car at the 2007 Los Angeles show, and for good reason: this will be the world's first production passenger car powered by hydrogen fuel cell, and is set to reach the dealers late in 2008. Though manufacturing numbers will be limited and the FCX will go to only a carefully selected group of approved pioneer customers in the United States and Japan, the technical achievement represented by the new vehicle is highly significant.

A development of the original FCX concept of 2005 as well as the Kiwami two years before that, the Clarity retains the vertical fuel-cell stack set in the central chassis spine of the vehicle: this was the breakthrough that gave the 2005 concept such good interior space. The nose is now fuller, the chromed bumper element weightier, and the wheels smaller, yet, thanks to a long wheelbase and a waistline that falls toward the front, the Clarity has a low-slung appearance; the subtle side feature line that rises through the front and rear door handles and drops again at the rear suggests comfort rather than performance. The hood and front overhang are longer than on the 2005 concept; this, and the substantial visual mass of the high tail, makes the Clarity appear more conventional than the concept, though by any standards it is a very futuristic vehicle.

The interior is stylish, modern, and high-tech looking, with the floating instrument pod not integrated into the main sweep of the dashboard and the center stack hanging from the main instrument panel leaving free space beneath. Yet with its multiple pods, control panels, and a Civic-like double-decker instrument display, the Clarity's driver environment has an air of complexity that is at odds with the smooth, gentle surfaces of the exterior.

Engine	Hydrogen fuel cell (Honda V stack)
Power	100 kW (135 bhp)

Honda Jazz/Fit

Engine	1.5 in-line 4
Power	88 kW (118 bhp) @ 5800 rpm
Torque	145 Nm (107 lb ft) @ 4800 rpm
Gearbox	5-speed manual
Installation	Front-engined/front-wheel drive
Front suspension	MacPherson strut
Rear suspension	Torsion beam
Brakes front/rear	Discs/discs
Front tires	185/55R16
Rear tires	185/55R16
Length	3899 mm (153.5 in.)
Width	1722 mm (67.8 in.)
Height	1524 mm (60 in.)
Wheelbase	2499 mm (98.4 in.)
Curb weight	1109 kg (2444 lb)
0–100 km/h (62 mph)	8.9 sec

Known as the Jazz in Europe and the Fit in North America and Japan, this small and versatile Honda hatchback has won a loyal following from the many who value its reliability, its economy, and the very clever way in which its rear seats fold to maximize cargo and passenger space. So it was something of a surprise when the second generation of this advanced design was revealed as a car that looked all but indistinguishable from its predecessor.

Yet what initially appears to be a disappointing spot-the-difference redesign proves on closer examination to have been a very skillful reshaping of the vehicle. Almost everything has been changed and improved, yet the overall impression remains the same – an important factor in preserving repeat-purchase loyalty and maintaining second-hand values.

Overall, the new car has a more substantial, solid look than its predecessor: surfaces have been filled out, the detailing strengthened, and lines and angles sharpened. The grille is now more angular and blends seamlessly with the hood and the windshield and roof to form a single arched profile. The headlamps and lower air intakes have grown in size, and the tall sides are relieved by surfacing that sees a mildly concave section below the top feature line blend with a second line emanating from the top of the rear wheel arch. An additional flute lower on the front door catches the light, again reducing visual mass, while above the waistline a quarter-light takes care of the area behind the screen pillar that is often awkward on monovolume vehicles.

At the rear the look is again similar, though the rear-window base now dips toward its center to accommodate the wiper spindle.

The largely circular theme of the interior and dashboard sees Honda employing less of a sure touch. The design looks somewhat cluttered and the surfacing is awkward in the area where the instrument housing meets the central navigation screen surround.

Honda Pilot

Sheltering at the Detroit show under a big blue sign saying "prototype," this 2008 design clearly advertised itself as a strong indicator of the size, shape, and proportions of this year's new Pilot, if not of every aspect of its detailing.

For some, it will be reassuring that Honda's biggest SUV has retained the cautious, sensible look of its well-respected predecessor; others will regret that so forward-looking a company as Honda has chosen not to move the game along, as it did in 2006 with the current CR-V.

Whatever the verdict on the Pilot's progressiveness, however, it clearly comes across as a strong, mature, and solid design. It is handsome in a calm, serene way, well proportioned and with strong but not aggressive feature lines. The surfacing around the windows makes them appear inset, giving the design a sense of security that is accentuated by the radiused corners to all the many windows. The impression is upright and unexciting and in some ways reminiscent of the previous CR-V, too, but its simplicity and strength will sell well in North America.

The only jarring note is that struck by the grille and the surrounding frontal treatment. The thick chrome surround seems an uncomfortably brash ambassador for the sober lines of the rest of the vehicle, and the overcomplex headlamps are likely to be seen as unappealing. Honda would do well to tone down these details before volume production.

The very pale interior color may not be representative of the production model, but it is likely that the basic architecture is. The look is perhaps not as exclusive as Pilot buyers might aspire to, but the unusual control layout, placing the transmission selector high on the dashboard, almost level with the steering wheel, liberates a large area of the center console for storage and—that essential for the US consumer— beverage holders.

Honda Puyo

Penned by Japan's most innovative carmaker, the Honda Puyo is in every sense a great example of the tradition of weird and wacky concept car designs that make the Tokyo show such a magnet for gadget freaks the world over.

Looking like the top half of a food blender set on four small wheels, the Puyo does not initially appear to have any clear purpose. The front, sides, and rear are all vertical; the four occupants peer out in all directions through the 360-degree glass dome superstructure; and the miniature wheels placed at each corner make it look as if the vehicle will tip over if driven too harshly. This could, perhaps, be a luxury golf cart or an electric shuttle for showing VIPs round a factory.

According to Honda, however, the mission of the Puyo is to embody "out of the box thinking," providing fun for both the vehicle owner and those around him or her. To this end, the Puyo boasts fuel-cell power and its tublike exterior incorporates a gel skin to make it soft and friendly to the touch, as well as, presumably, making it friendlier in a pedestrian impact. With no corners and no angles, the car is like a soft sculpture on wheels—almost as if it were a deflated version of the cube designs that were once fashionable in Japan.

Scissor doors open diagonally upward to provide access to the interior, which, at first glance, appears to have no controls or instruments whatsoever. It is only after a while that a small throttle pedal becomes evident and the control joystick is discovered nestling in a yellow panel on the driver's door; the displays appear (through a stretched cloth) only when the vehicle is started up.

This, like many other Tokyo concepts, is a solution to a problem few people knew existed. But that does not prevent the Puyo being amusing, original—and charmingly pointless.

Hummer HX

Design	Carl Zipfel
Engine	3.6 V6
Power	227 kW (304 bhp) @ 6300 rpm
Torque	370 Nm (273 lb ft) @ 5200 rpm
Gearbox	6-speed automatic
Installation	Front-engined/four-wheel drive
Front suspension	Short and long arm
Rear suspension	Semi-trailing arm
Brakes front/rear	Discs/discs
Front tires	20 in.
Rear tires	20 in.
Length	4343 mm (171 in.)
Width	2057 mm (81 in.)
Height	1829 mm (72 in.)
Wheelbase	2616 mm (103 in.)
Track front/rear	1702/1727 mm (67/68 in.)

Designed with the input of three graduates fresh from design college, Hummer's HX concept brings a more youthful and more adventurous take to the Hummer experience.

It is to the designers' great credit that the HX, though only less than 2 feet shorter than the H3 station wagon, looks much more sporty, more compact, and more ready for action. It immediately gives a remarkable impression of power and go-anywhere eagerness. Every design cue, from the big, exposed, tractor-like tires to the solid, sharply angled front skidplate and the huge wheel articulation, says this is a vehicle that can drive over—or through—any obstacle that appears in its path.

The rugged, military look extends to every area of the HX's design. Every panel, every corner appears heavily armored; bright-metal quick-release fasteners, highlighted against the olive-drab bodywork, secure the main panels to the structure in the manner of a tank or a troop carrier, and such important elements as the lights, exhausts, and air intakes are well protected. This looks like a vehicle that could be dropped by parachute from a helicopter and drive off undamaged.

The HX's extreme width and low, cut-down cab superstructure make for a squat, purposeful silhouette, while the outstretched wheels on the deliberately exposed suspension linkages suggest a tenacious grip on the surface underneath. The side profile is suitably purposeful, too, the cant-rail extensions bracing the triangle between the thick B-pillar and the tailgate to shift the visual center of gravity rearward and reinforce the strong look.

Opinions are more divided when it comes to the interior. Most of the visible areas—including the instrument panel, the instrument bezels, the transmission tunnel, and even the gear shifter—appear to be milled from solid metal, creating a shiny, hard, and cold look; the seats are thin and hard-looking, too, while the carpets have a camouflage pattern.

Hyundai Genesis

Design	Joel Piaskowski
Engine	4.6 V8 (3.3 and 3.8 V6 also offered)
Power	280 kW (375 bhp) @ 6500 rpm
Torque	440 Nm (324 lb ft) @ 3500 rpm
Gearbox	6-speed automatic
Installation	Front-engined/rear-wheel drive
Front suspension	Multi-link
Rear suspension	Multi-link
Brakes front/rear	Discs/discs
Front tires	225/40R19
Rear tires	245/40R19
Length	5024 mm (197 in.)
Width	1864 mm (73.4 in.)
Height	1491 mm (58.7 in.)
Wheelbase	2936 mm (115.6 in.)
Track front/rear	1575/1580 mm (62/62.2 in.)
Curb weight	1817 kg (4006 lb)
0–100 km/h (62 mph)	< 6 sec

Any Hyundai with Genesis in its title is a car to be taken very seriously indeed, for Genesis is the Korean carmaker's strategy for climbing upward into the premium segment and, probably sooner than most people expect, challenging the likes of BMW, Audi, and Lexus.

Underneath the skin are a V8 engine and a rear-wheel-drive chassis that promise to be as thoroughly engineered as anything out of Germany, and the vehicle's 16-foot length places it midway between Mercedes' E- and S-Class in stature. Yet skillful design, with smooth surfaces and flowing contours, gives the Genesis more of the air of a compact sports sedan than either of the two Mercedes. In terms of aesthetics the rear is possibly more successful than the front, which is more Toyota-like and lacks the strength and sophistication of European competitors.

Nevertheless, the executive feel is promoted by the provision of chrome trim on much of the exterior detailing, including the window frames, door handles, door sills, and the front and rear apron openings; in a classy feature, the chrome strip across the trunk lid extends into the clear-glass center strip within the rear lights.

The side-view proportions are those of a classic luxury sedan such, as a BMW 5 Series: the surface above the door handles winds along the full length of the body, connecting front and rear lights and emphasizing the length of the vehicle. On a practical level, the rear quarter-lights open with the doors, giving passengers easier access.

The production version was launched at the 2008 Detroit show eight months after the concept and displays bigger, bolder headlights and a stronger grille; this has the perhaps unintended side effect of making the car look slightly more regal and conventional. The interior has a smooth, sweeping dashboard that avoids overcomplication; a control knob similar to BMW's i-Drive device sits just behind the gear selector.

Hyundai Genesis Coupé

Design	Joel Piaskowski
Engine	3.8 V6 (2.0 in-line 4 also offered)
Power	228 kW (306 bhp) @ 6000 rpm
Torque	357 Nm (263 lb ft) @ 4700 rpm
Gearbox	6-speed automatic
Installation	Front-engined/rear-wheel drive
Front suspension	MacPherson strut
Rear suspension	Multi-link
Brakes front/rear	Discs/discs
Front tires	255/35ZR20
Rear tires	275/35ZR20
Length	4630 mm (182.3 in.)
Width	1864 mm (73.4 in.)
Height	1379 mm (54.3 in.)
Wheelbase	2819 mm (111 in.)
Curb weight	1610 kg (3549 lb)
0–100 km/h (62 mph)	< 6 sec
Top speed	240 km/h (149 mph)

The second weapon in Hyundai's strategy to move its brand upmarket to tempt premium customers (the first being the Genesis sedan), the Genesis Coupé is the ambitious Korean company's equivalent of the Audi A5 or the Mercedes-Benz CLK. Though derived from the engineering base of the Genesis sedan launched at the 2008 Detroit show, the Coupé concept nevertheless appeared two months earlier than the sedan—appropriately enough in the Californian environment of Los Angeles, where upscale coupés are big business.

Yet the aggressively body-equipped and wildly orange-and-black-colored LA concept did Hyundai no favors in establishing suitably classy preview imagery for the more restrained and much more tasteful production version, shown a few months later in New York. Especially interesting is the side surfacing, where a feature line originating from the front wheel arch rises to window level at the B-pillar; an equally sharp line runs forward from the rear deck, over the rear wheels, and into the door, where it undercuts the front crease to produce an unusual concave surface that starts just behind the wheel arch, broadens, and then twists to become near-horizontal as it goes rearward. Further interest is added by the distinctive step-down of the beltline as it travels from the door to the rear side window, producing a wave effect when viewed from the rear three-quarters.

The only slight disappointment in this interesting design is the nose treatment, where the slightly Volvo-like grille appears to make the bumper structure sag downward over the broad lower air intake.

While the longer, heavier Genesis sedan will be offered with a V8 engine (Hyundai's first), the Coupé will stick with a choice of two V6 units, driving the rear wheels via a six-speed automatic transmission.

Hyundai HED-5 i-Mode

Billed by Hyundai as a design study for a six-seater MPV, the HED-5 i-Mode is a vehicle that is middling in size but very high in complexity; so high, in fact, that its myriad styling details ruin whatever merits there once were in its basic overall shape. The designers have tried too hard, and the result is a vehicle that is muddled, confusing, and unattractive.

The underlying shape is a single sweeping arch flowing from the very fast windshield, along the cant rails, to the near-vertical rear panel. However, the excess of detailing fights against any simplicity: instead of conventional A-pillars, for example, Hyundai's designers have split the pillar into two for no apparent reason, marring the shape and impeding visibility. The DLO plays strange tricks at the rear, too, where the oval rear quarter-light drops below the waistline to follow the swirl of the cant rail as it sweeps round forward again. Only the smooth doors, with their plunging waistline, give relief from this swirling muddle, and even then they have a recessed section toward the rear.

Front and rear again supply an array of conflicting messages, resulting from the interplay of bulging convex surfaces and recessed and cut-in sections. The large headlights, set at either side of the clamshell hood, are displaced backward from the smaller lights below them: it looks like a bad mismatch, but it is clearly intentional.

The high-set rear is dominated by the big boomerang-shaped taillights that sweep down from the cant rail and swing forward into the HED's sides; the base of these lights forms the line of a further inset panel at the rear, just below the vertical portion of the rear glass.

We at *The Car Design Yearbook* often criticize carmakers for being too cautious in their designs, but the HED-5 proves something else: that trying every idea at once can be just as disastrous.

Engine	2.2 in-line 4 diesel
Power	158 kW (212 bhp)
Torque	461 Nm (340 lb ft)
Gearbox	6-speed automatic
Installation	Front-engined
Brakes front/rear	Discs/discs

Hyundai i-Blue

Design	Hyundai Japan
Engine	100 kW fuel-cell hybrid
Brakes front/rear	Discs/discs
Front tires	285/50R20
Rear tires	285/50R20
Length	4850 mm (191 in.)
Width	1850 mm (72.8 in.)
Height	1600 mm (63 in.)
Wheelbase	2850 mm (112.2 in.)

Given that the i-Blue is a clean-sheet-of-paper concept car powered by a hydrogen-fed fuel-cell hybrid powertrain, the Japan-based designers enjoyed an exceptional degree of freedom in deciding on the solution they would favor and the proportions the final design would have. Yet against the background of this unusual freedom the resultant design must be judged mildly disappointing, if only because by mixing so many design messages the designers have missed the opportunity to convey on a visual level the excitement and futuristic nature of the car's zero-emission powertrain.

The i-Blue's format is that of an SUV crossover, imbued with a deliberately strong dose of dynamism in its external design in order to disguise its height and, to a lesser extent, its length. Thus the pointed front and tapering rear give it an aggressive air, an impression strengthened by the very fast windshield—which would pose visibility problems for the driver—and the plunging side-window baseline with its jagged step, formed by the door handles, just aft of the B-pillar.

When the car is viewed from the side, the proportions look unbalanced, with the high waistline, the weighty rear roof section, and the peculiar triangular DLO profile accentuating the top-heavy feel. A chromed slash recessed into the lower doors does little to relieve the mass of the sides, though the solution around the rear lights is more elegant.

Inside, Hyundai's designers have clearly taken inspiration from science-fiction movies, with steeply reclined seats and a spaceship-style steering module incorporating touch-screen control pads and twin handgrips. Instruments, displays, and read-outs, most of them in a shade of electric blue, are arranged in two tiers in front of the driver, on the side of the center console and in a full-width band across the dashboard. Yet while the intention may be futuristic, the execution is perhaps too intimidating to be taken seriously as a practical production-car proposition.

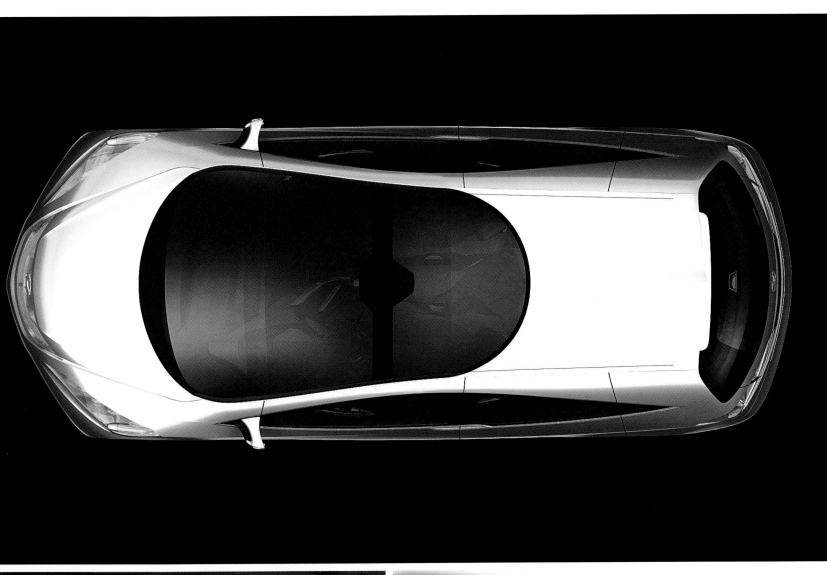

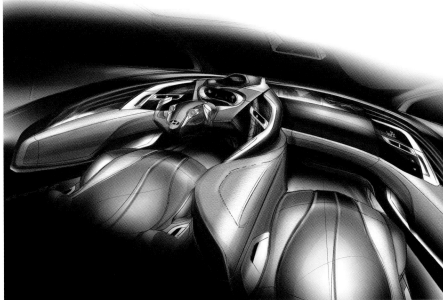

Hyundai Veloster

Design	Suk-Geun
Engine	2.0 in-line 4
Gearbox	5-speed automatic
Front tires	245/35R20
Rear tires	245/35R20
Length	4100 mm (161.4 in.)
Width	1790 mm (70.5 in.)
Height	1450 mm (57.1 in.)
Wheelbase	2600 mm (102.4 in.)

Think of this as Hyundai's equivalent of the Audi TT—or at least the previous-generation, more compact TT—and you will not be far off target. The Veloster, its name a compound of *velocity* and *roadster*, is tasked with putting the Korean company on the wish list of the well-heeled Y-generation buyer in his—or even her—twenties, and as a Seoul show concept it attracted a lot of attention and positive comment.

Like the TT, the Veloster is a single-minded design with a single-minded message: sportiness. From every angle it is dominated by its huge 20-inch wheels and the bulging wheel arches that house them, while viewed from the side the profile is deliberately ambiguous: the overhangs are very short, the windshield is thrown forward so its base is almost at the center line of the front wheels, and the roofline is kept high toward the rear, leaving the observer wondering whether it is front- or mid-engined.

The stance is squat, purposeful, and meaty; the low-set grille mouth appears to be eager to scoop in large gulps of air; and the substantial sills visually help the car to hug the ground. The massive, vanlike rear pillar is another source of ambiguity: it disguises the configuration of cabin and load areas and, with its solidity and the slope of the tail panel, provides a propulsive energy to the design's profile.

On a more detailed level it is clear that Hyundai's designers have left the front end—and especially the hood—clean and simple in order to focus attention on the relationship of the body to the ground; the wheel spokes subtly pick up the red body color, and this is highlighted, too, in a smart interior that bravely combines red, white, and metallic surfaces. While the effect might be more flashy than sporty, it is nonetheless novel, and attractive.

Infiniti EX

Design	Nissan Technical Center
Engine	3.7 V6
Power	261 kW (350 bhp) @ 6800 rpm
Torque	350 Nm (258 lb ft) @ 4800 rpm
Gearbox	7-speed automatic
Installation	Front-engined/all-wheel drive
Front suspension	Double wishbone
Rear suspension	Multi-link
Brakes front/rear	Discs/discs
Front tires	225/60R17
Rear tires	225/60R17
Length	4631 mm (182.3 in.)
Width	1803 mm (71 in.)
Height	1589 mm (62.6 in.)
Wheelbase	2799 mm (110.2 in.)
Curb weight	1776 kg (3915 lb)
Fuel consumption	11.8 l/100 km (19.9 mpg)

Infiniti, the premium division of Nissan, is just one of many carmakers seeking to soften the image of its SUVs so as to prevent customers from switching to other styles of vehicle that they perceive as being less aggressive and less fuel-hungry and therefore more socially acceptable.

In the EX, Infiniti's designers have cleverly exploited the brand's trademark soft, flowing style to bring a gentler and more accommodating feel to this four-wheel-drive crossover line. In particular, the high-riding blockiness of traditional SUVs has been completely eliminated: despite the big wheels, the black wheel arch surrounds, and the noticeable ground clearance, the EX's stance has something of a coupé quality to it. Indeed, Infiniti brands the EX as the world's first compact crossover coupé.

Much of this stems from the small and subtle Infiniti grille, which leads the eye toward the L-shaped headlights that flow—somewhat awkwardly—into the fenders. The hood feature line running rearward from the grille is drawn back to meet the door mirror, where it cleverly splits the airflow above and below the mirror. At the side, the subtle feature line from the rear tip of the headlight dips before rising again over the back wheel arch, again breaking up the mass of the body side; this, in conjunction with the arched DLO, makes the car look slimmer and much lower than it actually is.

The interior of the close-to-production concept shown at New York in 2007 was typically show-car in its use of white leather for almost every surface. However, in terms of its architecture and its equipment it directly anticipated the production model, where the finishes are equally impressive but the colors more conventional. A stand-out feature is an around-view system, wherein cameras mounted at front and rear and on each side feed into the navigation display monitor to give the driver a bird's-eye view of the vehicle to aid parking in tight spaces.

Infiniti FX

Design	Shiro Nakamura
Engine	5.0 V8
Power	291 kW (390 bhp)
Torque	500 Nm (368 lb ft)
Gearbox	7-speed automatic
Installation	Front-engined/all-wheel drive
Front suspension	Double wishbone
Rear suspension	Multi-link
Brakes front/rear	Discs/discs
Front tires	265/45R21
Rear tires	265/45R21
Length	4803 mm (189.1 in.)
Width	1925 mm (75.8 in.)
Wheelbase	2850 mm (112.2 in.)

Infiniti, Nissan's premium brand, is launching into Europe in 2008—and it is a measure of the importance the Renault–Nissan group attaches to Europe that it chose to hold the world premiere of the second-generation FX not in the US, but at the Geneva Motor Show.

In the twenty years that Infiniti has been in existence, all its launches have been at North American shows, and all its sales have been concentrated there and in the Middle East and Russia. The move to Europe, where competitiveness in the luxury-car market is tougher than anywhere else in the world, is a psychologically important one for Infiniti.

So it comes as a surprise that the model Infiniti has chosen to celebrate its European debut is a type that is deeply out of favor among CO_2-conscious commentators in Europe: a heavyweight, high-powered luxury SUV. Infiniti points out that political opposition to SUVs has not dented buyers' enthusiasm for them, and that the sector continues to flourish.

Infiniti also notes that the FX hardly conforms to the conventional stereotype of a big SUV: it may be powered by a 5-liter V8 engine, but it is sensitively shaped, with a fluid, racy look rather than a Rambo blockiness, and it has sports-car proportions and is built on a sports-car platform.

The large, horizontally oriented grille certainly bears this out, as do the flowing lines of the front fenders and the sweeping curve of the DLO. Air outlets behind the front wheel arches add to the effect. At the rear, the coupé effect is again apparent, the only discordant element being the chromed roof rails.

Similarities with the smaller EX are strong; from the rear it takes a trained eye to tell them apart. Inside, the sporty feel is continued, with a small steering wheel, a cockpit-like driving position—though the driver sits high—and a neat recessed navigation screen with the keypad in front of it.

Infiniti G37

Engine	3.7 V6
Power	443 kW (330 bhp) @ 7000 rpm
Torque	366 Nm (270 lb ft) @ 5200 rpm
Gearbox	6-speed manual
Installation	Front-engined/rear-wheel drive
Front suspension	Double wishbone
Rear suspension	Multi-link
Brakes front/rear	Discs/discs
Front tires	225/45R19
Rear tires	245/40R19
Length	4651 mm (183.1 in.)
Width	1824 mm (71.8 in.)
Height	1395 mm (54.9 in.)
Wheelbase	2850 mm (112.2 in.)
Track front/rear	1544/1560 mm (60.8/61.4 in.)
Curb weight	1664 kg (3668 lb)
Fuel consumption	10.7 l/100 km (22 mpg)

Longer, lower, and wider than its G35 predecessor, the new G37 brings Infiniti's big coupé offering up to date and incorporates much of the look and feel of the well-received concept that appeared at the 2006 Detroit show and that we reviewed in *The Car Design Yearbook 5*.

In moving from concept to production status, the design has lost some of its smoothness and fluidity, though its forward-leaning stance is similar, and such key details as the front and rear lights have similar themes to their shaping, if not exactly the same proportions.

The signature Infiniti grille is more distinct than on the concept, where its horizontal bars merged into the body paneling at either side; the grille leads the eyes out to the headlights, which sit below the raised edges that are carried the full length of the hood to the base of the A-pillar. The strong curves at the front are echoed in the gently flared wheel arches and the curve of the roof surface, while the waistline rises steadily toward the rear to concentrate the visual center of mass around the rear wheels. The cabin wraps in toward the rear, too, increasing the prominence of the rear fenders and, along with the steeply raked rear window, imparting a strong propulsive dimension to the design.

At the rear, large L-shaped lights echo the shape of the headlights and convey a smiling impression; unfortunately, this is spoiled by the awkward placing of the high-level brake light right on the apex of the trunk lid.

The interior, a combination of black and silver for the controls and switches and gray for the comfort and convenience features, is effective enough but lacks the distinctiveness one might hope for in a high-end coupé—unless, of course, violet electroluminescent instrumentation and aluminum trim inspired by handmade Japanese Washi paper are on the buyer's list of priorities.

Italdesign Quaranta

Design	Giorgio Giugiaro
Engine	3.3 V6 plus electric hybrid drive motors
Power	200 kW (268 bhp) @ 5500 rpm
Torque	288 Nm (212 lb ft) @ 4400 rpm
Gearbox	None
Installation	Mid-engined/all-wheel drive via wheel motors
Front suspension	Double wishbone
Rear suspension	Double wishbone
Brakes front/rear	Discs/discs
Front tires	275/35R20
Rear tires	275/40R20
Length	4450 mm (175.2 in.)
Width	1982 mm (78 in.)
Height	1230 mm (48.4 in.)
Wheelbase	2620 mm (103.1 in.)
Track front/rear	1759/1697 mm (69.3/66.8 in.)
0–100 km/h (62 mph)	4.05 sec
Top speed	250 km/h (155 mph) limited
Fuel consumption	7.1 l/100 km (33.1 mpg)
CO$_2$ emissions	180 g/km

Created to celebrate the fortieth anniversary of Giugiaro's highly successful Italdesign vehicle design and development business, the Quaranta—Italian for "forty"—is a grand supersports concept of the type that Giugiaro has habitually done so well.

Like the McLaren F1 of the 1990s, the Quaranta is mid-engined and places its driver in the center, with the two passengers sitting outboard slightly further aft; there is also provision for a small child seat behind the driver. The doors and roof rise to allow access to the interior.

Giugiaro bills the Quaranta as an "extreme" environment-friendly supercar: its Toyota-derived engine and hybrid system drives all four wheels via electric motors rather than a conventional transmission.

The idea of the project was also to reprise developments pioneered on earlier Italdesign concepts: hence the collaboration with Toyota on the powertrain, the multicontrol steering-wheel system from the 1980 Medusa, and suspension that uses the central monoshock principle first seen on the 2004 Volta.

Visually, too, the Quaranta checks off all the boxes expected of a Giugiaro design: clear, flowing lines, crisp surfaces, edges, and angles, and generally restrained detailing. Yet there is something missing.

The overall shape, heavily cab-forward in its proportion, does not work well. Giugiaro's idea of having the full-width continuous glazing running from the front spoiler, over the windshield and roof and back to the edge of the tail, does not work either; the built-in solar cells may provide useful power, but the black band breaks up the shape of the car, preventing it from showing harmony or balance. The rear end, too, is disappointing: it is an inelegant flat panel flanked by large triangular air outlets with the taillights sunk into their upper edges.

Giugiaro is not known as the Maestro for nothing. He and his team have produced dozens of iconic, forward-looking designs over the past forty years. It is a shame that the design produced to celebrate this success has lost that momentum.

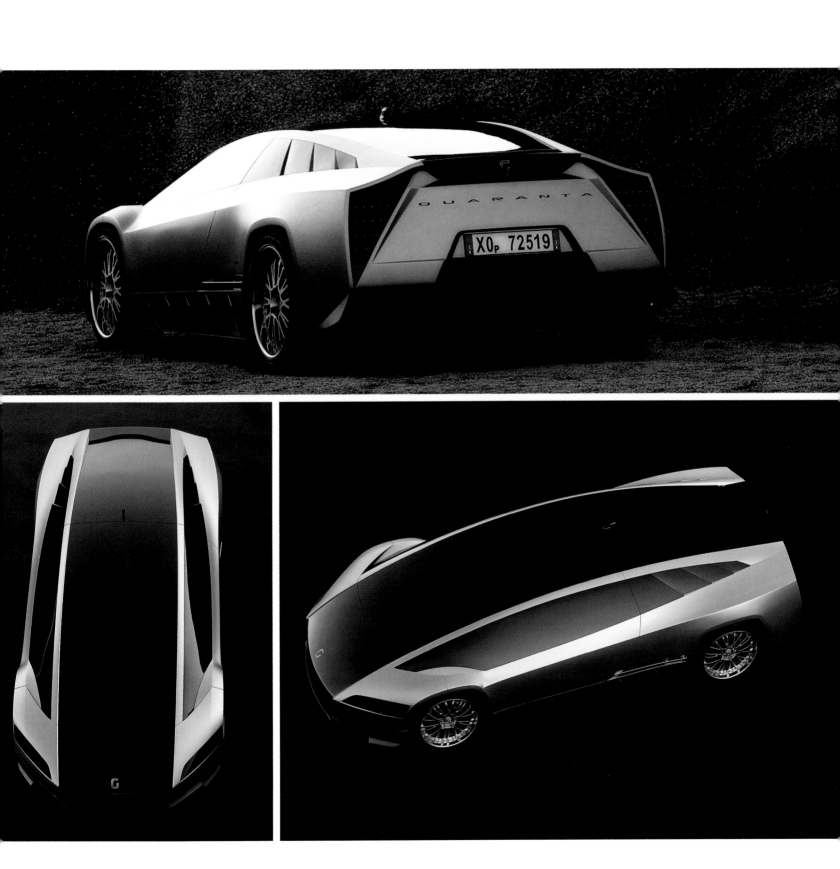

Jaguar XF

Design	Ian Callum
Engine	4.2 V8 (2.7 V6 diesel and 3.0 V6 gasoline also offered)
Power	306 kW (410 bhp) @ 6250 rpm
Torque	560 Nm (413 lb ft) @ 3500 rpm
Gearbox	6-speed automatic
Installation	Front-engined/rear-wheel drive
Front suspension	Double wishbone
Rear suspension	Multi-link
Brakes front/rear	Discs/discs
Front tires	20 in.
Rear tires	20 in.
Length	4961 mm (195.3 in.)
Width	1877 mm (73.9 in.)
Height	1460 mm (57.5 in.)
Wheelbase	2909 mm (114.5 in.)
Track front/rear	1559/1571 mm (61.4/61.9 in.)
Curb weight	1842 kg (4061 lb)
0–100 km/h (62 mph)	5.1 sec
Top speed	250 km/h (155 mph)
Fuel consumption	12.6 l/100 km (18.7 mpg)
CO$_2$ emissions	299 g/km

The XF is the production version of the C-XF concept that graced the cover of *The Car Design Yearbook 6* and counts as perhaps the most eagerly awaited new model in Jaguar's seventy-five-year history. The mission of the concept was to fast-forward Jaguar design into the twenty-first century; this it did with some aplomb, putting Jaguar back onto the style map where it always used to belong.

But now the production XF needs also to capture the public and the financial community's imagination so that confidence is maintained in the Jaguar brand following the long-drawn-out and psychologically damaging process that led to the sell-off, for US$2.3 billion, to India's Tata Motors in March 2008. To achieve this, the XF needs to be a major step forward.

In Jaguar terms the XF is indeed a major leap into contemporary style: gone, convincingly, is the old-world look the brand used to inhabit. The new face of Jaguar has large, catlike headlamp lenses that draw the eye outward from the simple rectangular mesh grille; the carefully contoured hood creates a tantalizing interplay of light and shadow; and the strongly coupé-like lines proudly advertise Jaguar's new position as a producer of sports cars—some of which happen to seat five people.

The coupé profile is further accentuated by the chrome band encircling the side windows, while the tail is an especially elegant piece of design, much less fussy than on the concept. Where the production car loses out to the concept is at the front: the concept's deeply recessed grille and less bug-eyed lights gave it greater visual impact.

The XF is a car that does not always look as good in photographs as in the metal; the same holds true for the wood, metal, and leather environment of the interior. The dashboard is a masterpiece of calm and assured high-quality design, combining simple style with intuitive ergonomics. As a template for Jaguar's new future, it is an encouraging start.

Jeep Renegade

Design	Tony Shamenkov
Engine	Twin 200 kW (268 bhp) electric motors plus 1.5-liter 3-cylinder diesel engine as range extender
Gearbox	None
Installation	Front diesel engine, four-wheel drive through electric motor on each axle
Brakes front/rear	Discs/discs and regenerative braking
Length	3885 mm (153 in.)
Width	1598 mm (62.9 in.)
Height	1431 mm (56.3 in.)
Wheelbase	2580 mm (101.6 in.)
Fuel consumption	Potential max. 2 l/100 km (110 mpg)

For as long as the iconic back-to-basics Jeep has existed—more than sixty years—generations of designers have been trying to re-invent it. Jeep's own designers have contributed their fair share of concept studies at successive Detroit shows; some of these have influenced subsequent production models, while others have proved to be dead ends in terms of product progression. All, however, have somehow managed to encapsulate the rugged, go-anywhere spirit of the brand.

The 2008 Renegade concept, however, is something different. Not only is its fundamental propulsion system electrical (the diesel engine is just a back-up), but it also breaks with many of the signature Jeep design rules that, to date, have made Jeeps always instantly identifiable as Jeeps.

Perhaps foremost among these characteristics has been the blocky, straight-lined look, built up of flat or substantially flat panels. The Renegade's open two-seater body is built up of much more complex surfaces, many of which incorporate soft curves in their shaping, and the whole vehicle looks as if it has been conceived using a flexible template rather than a right-angled one.

The pared-down dune-buggy format of the concept is a new one for Jeep, but the design team has successfully imbued it with a sense of minimalist fun and involvement with the surrounding elements. The windshield is minimal, no provision is made for a roof, and the doors have large cutaways; inside, the style is more like that of a small sports car than of a rugged 4x4.

Huge tires lessen the damage caused to sensitive terrain and add to the Renegade's chunky appearance, while the payload deck behind the two occupants can house anything from mountain bikes or diving gear to the twin water-scooters of the show car. Interior innovations include novel construction methods and an instrument panel that uses wireless rather than cable connections.

While the Renegade may be very different, a measure of its success is that it is still unmistakably a Jeep.

Ken Okuyama K.07 & K.08

Details given are for the K.07

Design	Ken Okuyama
Engine	2.0 in-line 4
Power	179 kW (240 bhp) @ 8300 rpm
Torque	206 Nm (152 lb ft) @ 7000 rpm
Installation	Mid-engined/rear-wheel drive
Front suspension	Double wishbone
Rear suspension	Double wishbone
Brakes front/rear	Discs/discs
Length	3580 mm (140.9 in.)
Width	1850 mm (72.8 in.)
Height	1100 mm (43.3 in.)
Curb weight	750 kg (1654 lb)

Ken Okuyama is a well-known figure in auto-industry design circles, having worked in the studios of GM, Porsche and, most recently, Pininfarina in Italy. But while many car designers are content with being associated with well-respected brands, for others, such as Okuyama, having a car with their own name on it is a career objective.

For Okuyama, the moment came in March 2008 when, at the Geneva Motor Show, he unveiled his first two designs—the K.07 and K.08. Both are ultralightweight high-performance sports cars in the vein of familiar Lotus designs as well as more recent arrivals on the market.

The stark, open-wheeled simplicity of the K.07 is reminiscent of the Caterham models, themselves derived from the original Lotus Seven. However, the Okuyama design is very smooth and slick, with a pointed nose leading to a small, rectangular grille, and the engine is rear- rather than front-mounted. The cockpit is fully open, with no windshield; instead, twin fairings shield the driver's instruments and the passenger-side dials, and strong rollover towers protect the occupants in the event of an accident. The rear design is particularly neat—in contrast to the complex treatment of the more obviously aerodynamic KTM X-Bow (*The Car Design Yearbook 6*)—and the aluminum and carbon-fiber structural materials are left unpainted. Inside, there has been no attempt to add comfort or convenience, but the materials and finishes—especially the TAG Heuer instruments—are of the highest quality.

The K.08, meanwhile, is a compact closed coupé, much in the style of the Lotus Exige; it uses the same running gear as the K.07 but has fully enclosed wheels and suspension. Its sunglasses-like window line is powerfully symbolic, and the complex array of LED headlight elements, which run both vertically on the inside edge of the front wheel housings and horizontally along the bridging element to the main nose nacelle, are a reminder of the design's open-wheeled origins.

Kia Kee

Design	Peter Schreyer
Engine	2.7 V6
Power	149 kW (200 bhp)
Gearbox	6-speed automatic
Brakes front/rear	Discs/discs
Front tires	245/40R20
Rear tires	245/40R20
Length	4325 mm (170.3 in.)
Width	1860 mm (73.2 in.)
Height	1315 mm (51.8 in.)
Wheelbase	2675 mm (105.3 in.)

Kia wastes no time in explaining the significance of the name "Kee" for its latest concept car, designed under Peter Schreyer, architect of many notable models, including the iconic Audi TT. The Kee, says Schreyer, is genuinely key to Kia's future, containing as it does ideas that could be incorporated into future Kia models and help to build up an identity for a brand that does not yet have a heritage.

The citrus-yellow concept comes across as a compact and sporty ground-hugging two-seater sports coupé, though in fact it seats four: rear passengers nestle between the thick rear pillars, which are clad with black bands to link the wraparound rear window with the side glass. The emphasis throughout is on strong proportion rather than details; one of the few feature lines is the modestly kicked-out sill that travels rearward before turning upward at the rear wheel arch to meet the subtle shoulder-line crease.

The frontal treatment is unusual and perhaps less successful: headlamps of complex shape fill the space scooped out between the plain, chrome-lined grille opening and the front apron, with C-shaped chrome strips outlining the perimeter and cutting back into the center halfway up the headlights.

The rear is severe, but certainly distinctive: the neat lights invert the style of those at the front, flicking toward the center to form a well-defined tailgate lip, and the dark, narrow slot of the wraparound rear window gives a brooding impression beneath the heavily crowned roof.

The interior plays on shades of gray in different textures; the main dashboard slopes strongly forward and is matt-finished—in strong contrast to the center console, which appears to be a solid aluminum casting, polished to a high mirror finish. Novel toggle switches on the center of the dash echo this style in a rather less gaudy fashion.

Kia Kee **Concept** 139

Kia KND

Design	Kia Namyang Design Center
Engine	2.2-liter turbo diesel
Installation	Front-engined/all-wheel drive
Length	4466 mm (175.8 in.)
Width	1860 mm (73.2 in.)
Height	1600 mm (63 in.)
Wheelbase	2630 mm (103.5 in.)

For all its luminous lime-green paintwork and sharp, aggressive detailing, this compact SUV crossover is a likely successor to the mild-mannered Kia Sportage, the ambitious Korean company's current contender in the fashionable market for family-sized cars with a flavor of off-road adventure.

Kia's intention is that this vehicle should look good in both an urban and a rural environment; its design is undoubtedly a mixture of many themes, resulting in certain conflicts, though these might be eased somewhat with the switch to a less strident paint color.

The overall impression is of a pointy, edgy design: the wide black "mouth" formed by the high-set grille and headlights tapers to sharp points on the front fenders; the windshield is set at a very fast angle, giving a sharp leading edge to the side window, while the arched DLO drops obliquely to the rear, making for a sharp point at the trailing edge of the rear side window. That same slope leaves a large mass at the top of the C-pillar, much in the style of the Renault Mégane three-door; yet, in contrast to the Renault, the pillar is very thick, leaving a top-heavy mass of roof to meet the near-vertical rear window. The effect is similar to that of Nissan's 2007 Bevel concept.

The darkened-glass rear windshield and the oddly angled rear lights blend together in an inverted U-shape, which is reflected in the raised center section of the bumper moulding. Large wheels, high ground clearance, the narrow DLO, and the deep body sides do little to relieve the bulky feel of the design, even though the engine is only 2.2 liters and the car just 14 feet, 8 inches in length.

Two long doors give access to a predominantly white, four-seater cabin, which, with its numerous separate modules and zones, appears to lack a coherent design theme.

Kia Koup

Design	Peter Schreyer
Engine	2.0 in-line 4
Power	216 kW (290 bhp)
Torque	392 Nm (289 lb ft) @ 2000–4000 rpm
Gearbox	6-speed automatic
Installation	Front-engined/front-wheel drive
Front suspension	MacPherson strut
Rear suspension	Dual-link
Brakes front/rear	Discs/discs
Front tires	245/40ZR19
Rear tires	245/40ZR19
Length	4485 mm (176.6 in.)
Width	1845 mm (72.6 in.)
Height	1428 mm (56.2 in.)
Wheelbase	2650 mm (104.3 in.)
Track front/rear	1536/1536 mm (60.5/60.5 in.)

Kia, like its Korean owner Hyundai and like Mazda of Japan, is clearly a company on the move: its prodigious output of motor-show concept vehicles points to a big rush of additional production models in a few years' time.

The role of the Koup, displayed at the 2008 New York show, is, says Kia, to preview themes for a sedan model due in two years' time. This in turn indicates a very quick development process, the time between the signing-off of the exterior design and the start of volume production normally being between two and three years.

Clean, simple, and well resolved, the Koup is a four-seater coupé with fresh, clear design cues that would translate well into a sedan format. The front end is kept simple, marking the Koup as a mid-market car, but framed air scoops set into the outer edges of the underbumper area lend a serious, sporting touch. The sides are again cleanly surfaced, with noticeable wheel-arch flares and a kick-up of the sill feature behind the trailing edge of the door. The waistline drops at the A-pillar and rises again gradually toward the rear, where the design is particularly attractive. There is a crisp, gentle curve to the trunk spoiler, and the neat taillights and the Kia emblem sit in the scooped-out vertical panel beneath. Below bumper level, the black-finished diffuser-style inset panel carries the twin exhaust tailpipes and rises in the center to provide the license-plate mounting.

The well-ordered design of the interior shows, again, that this is a near-production car. The sloping dashboard places three aluminum-rimmed instruments directly in front of the driver, the aluminum theme being echoed in the center stack and the gear-lever surround on the center console. However, the black-and-white patterning of the seats is not likely to make it through to production.

Kia Soul

Design	Peter Schreyer
Front tires	245/40R19
Rear tires	245/40R19
Length	4105 mm (161.6 in.)
Width	1785 mm (70.3 in.)
Height	1610 mm (63.4 in.)
Wheelbase	2550 mm (100.4 in.)

Displayed at the 2008 Geneva show was the near-production version of the Kia Soul, a concept first aired at the Detroit show in January 2006. That concept was praised on several counts: for its tough, chunky attitude, for its sense of adventure, and for its trendy urban-chic looks. This version appears to all intents and purposes identical to the previous concept, the main technical difference being the inclusion of four conventional side doors instead of the earlier version's system of rear-hinged back doors.

Unusually, Kia has now chosen to show the Soul in three different versions, differing in trim, color, and equipment, and aimed at three distinct buyer groups. The Soul Diva is targeted at the modern fashion-conscious woman and is finished in white, with gold external detailing, including the wheels, door handles, and roof rails. Inside, the fabrics and surfaces are highly tactile. The Soul Burner is just the opposite: finished in black and adorned with dragon tattoos externally, it has extra lights and a fancy exhaust layout, while inside, its four sports seats and the surrounding upholstery are red and black. The Soul Searcher, finally, is intended as a zone of tranquility for people who like to take things easy; silver-gray in color, it has a full-length fabric sunroof offering a near-convertible effect. It is more subdued and discreet in its presentation and is perhaps the most likely prospect as a mainstream production model.

In terms of interior design, this near-production model keeps a similar overall architecture to the original concept but dispenses with some of its more offbeat ideas, such as the laptop computer that swings out in front of the passenger, and the circular CD player mounted against the head lining.

Up to date technically and likely to be produced at Kia's European factory, the Soul could prove to be a plausible competitor against such models as the Skoda Yeti, the Nissan Note, and even the upcoming Toyota Urban Cruiser.

Lamborghini Reventón

Engine	V12
Power	485 kW (650 bhp) @ 8000 rpm
Torque	660 Nm (486 lb ft) @ 6000 rpm
Gearbox	6-speed manual
Installation	Mid-engined/all-wheel drive
Front suspension	Double wishbone
Rear suspension	Double wishbone
Brakes front/rear	Discs/discs
Front tires	245/35ZR18
Rear tires	335/30ZR18
Length	4700 mm (185 in.)
Width	2058 mm (81 in.)
Height	1135 mm (44.7 in.)
Wheelbase	2665 mm (104.9 in.)
Track front/rear	1635/1695 mm (63.4/66.7 in.)
Curb weight	1665 kg (3671 lb)
0–100 km/h (62 mph)	3.4 sec
Top speed	340 km/h (211 mph)
Fuel consumption	21.3 l/100 km (11 mpg)
CO_2 emissions	495 g/km

Strictly speaking, the Reventón is a partially rebodied and even more imposing derivative of Lamborghini's top Murciélago LP640 model, if such a thing can be imagined.

The Murciélago has always been fearsome and formidable, albeit in a sophisticated and classy way. The Reventón shatters any claim to subtlety or discretion, with a suite of aggressive carbon-fiber body add-ons that turn it into a truly terrifying proposition. The style throughout is that of the sinister sharp points, jagged edges, and skewed planes characterizing secretive stealth aircraft; the menacing militaristic radar-reflective gray of the exterior compounds the scary effect.

It could of course be argued that this exaggerated style represents a return to the values of a classic Lamborghini: the legendary Countach, designed by Bertone's Marcello Gandini, went to controversial extremes with its outrageous multiplanar shape, especially at the rear, where huge air scoops, grilles, and outlets dominated each side. The Reventón picks up precisely the same jet-fighter exhaust theme with its massive five-sided rear grilles.

Aesthetics apart, the other extraordinary aspect of the Reventón is its price. Cast in the mold of a rare collector's piece, the Reventón will be produced in only twenty units—at a million euros apiece, or four times the price of the standard car. The fact that all the units were instantly sold says something about the desire of collectors to possess something unique, never mind the logic.

Art, engineering, or exhibitionism? A genuine technical advance or a not-so-subtle marketing ploy? Aesthetic improvement or an overblown bodykit that, if applied by a German tuning firm, would be condemned as sacrilegious? *The Car Design Yearbook* leaves it for others to decide.

Lancia Delta

Design	Lancia Style Centre
Engine	1.8 in-line 4 diesel (1.6 diesel, and 1.4, 1.9, and 2.0 gasoline, also offered)
Power	149 kW (200 bhp)
Gearbox	6-speed manual
Installation	Front-engined/front-wheel drive
Brakes front/rear	Discs/discs
Length	4500 mm (177.2 in.)
Width	1800 mm (70.9 in.)
Height	1500 mm (59.1 in.)
Wheelbase	2700 mm (106.3 in.)

The new Delta has been one of the most eagerly awaited Lancias for many years. The Lancia brand—a subsidiary of Fiat—has been struggling with declining, or at best stagnant, sales for many years, and its recent models have failed to capture attention outside Italy. Yet there is tremendous goodwill surrounding the make, relating in particular to the Delta hatchback, which was a big success in the 1980s and which became a legend on the rally circuit, too.

So when Lancia showed a flamboyant concept car, labeled Delta HPE (blending in another emotive name from the past), at the 2006 Paris show, the excitement became tangible. Few believed that the more extravagant features of the very attractive concept could be translated into the production car. Yet if anything the personality of the design has been strengthened, and the unusual proportions and stance of the car—especially at the rear—will ensure that it is a distinctive sight on the street.

First impressions are of a car that is surprisingly long for a medium hatchback; much of this impression is down to stylistic touches, such as the darkened cant rail (the dark roof wraps round to the upper window line), the dark sills, and the exaggerated forward plunge of the waistline. A crease line runs the full length of the side, emanating from the row of LEDs along the base of the headlight. The crease sharpens toward the rear to curve round and underscore the forward-sloping rear lights. The waistline rolls up toward the C-pillar to a triangular point, giving the intriguing impression that the roof is supported on a single point.

The large glass upper tailgate is voluminous and connects to a simple metal inset plinth that carries the number plate, leaving the glass overhanging. The only slight disappointment in a car that in many ways is a concept car made real is the interior, which lacks the originality and sense of style that are expected from Lancia.

Land Rover LRX

Design	Gerry McGovern
Engine	2.0 in-line 4 diesel with hybrid electric power
Installation	Front-engined/four-wheel drive
Brakes front/rear	Discs/discs
Front tires	20 in.
Rear tires	20 in.
Length	4351 mm (171.3 in.)
Width	1895 mm (74.6 in.)
Height	1535 mm (60.4 in.)
Wheelbase	2660 mm (104.7 in.)
CO_2 emissions	Potential 120 g/km

The LRX concept signals big changes at Land Rover, and at many levels, too. In years to come, it may be seen as a key turning point in the brand's development. Purely coincidentally, of course, it was presented just days after the announcement that India's Tata was the front-runner to buy the Land Rover and Jaguar makes from Ford.

On a strategic level, the compact LRX shows Land Rover moving toward a younger, sportier market and at the same time beginning to balance the high CO_2 emissions of its large models with a small and environmentally responsible design. This is essential if the brand is to have an independent future in such markets as Europe, where carmakers may be taxed on the fleet average CO_2 emissions of the models they produce.

On a design level, the LRX bears the first hallmarks of the aesthetic strategy of Land Rover's new design director, Gerry McGovern. The look is sharp, fresh, and sporting, the stance broad and aggressively poised; yet, thanks to the floating roof, the Range Rover-like inclination to the tail, and the clamshell shut to the hood, it is still very clearly a Land Rover. Front and rear overhangs are minimized to make the package look tight and to accentuate its off-road feel, yet from the side the most striking impression is that of the exaggerated tapering of the glasshouse toward the rear.

Detailed innovations abound, especially in the interior, where McGovern's team has used a futuristic mix of classic materials—leather, milled metal—and modern plastics and lighting technologies. Highlights are the soft, vegetable-tanned hide seats, the multilayer instrument graphics, and the ingenious mood illumination, which changes the color of the instruments, controls, and cabin background according to the type of driving—with red for sport, green for eco, and blue for general driving.

This is very much a miniature version of the highly successful Range Rover Sport, but with a much-reduced carbon footprint. Marketed correctly, it will have the capacity to create a fashionable—and profitable—new premium sector, just like the Mini did.

Lexus LF-Xh

Engine	V6 and electric motor hybrid
Installation	Front-engined/front-wheel drive; electric motor to rear axle
Brakes front/rear	Discs/discs
Length	4800 mm (189 in.)
Width	1895 mm (74.6 in.)
Height	1650 mm (65 in.)
Wheelbase	2850 mm (112.2 in.)

This new Lexus design study is a pointer toward the replacement for the current (and very successful) RX luxury sports SUV, and employs a hybrid drivetrain. The latter is no surprise, since Lexus, as Toyota's premium nameplate, has made it a central plank of its brand ethos to include an emissions-reducing hybrid drivetrain in all its top models. The surprise comes in the extreme fashion in which Lexus has reacted to the arrival of such sporty, coupé-profile SUVs as the BMW X6 and the Range Rover Sport. It could even be said that in some ways the LF-Xh looks more like the handiwork of a California customizing shop than of Lexus's own design department; the exaggerated rake of the windshield and the very narrow windows combine with the still-deep body sides to produce the low-rider look beloved of the cruising set.

A straight feature line runs from the front to the rear lights, separating the two contrasting styles of body architecture: the lower mass, which has the big wheels, high ground clearance, and tough stance of an off-roader, and the lighter upper structure, which is more akin to that of a racy coupé.

It is a blend that continues to provoke mixed reactions, especially as the design's front gives very different impressions depending on the angle of viewing. From below the short overhang, tucked-in understructure, and overbite grille look awkward, while from above the inward sweep of the back-set headlights and the projecting grille give an almost retro effect of a separate hood and fenders.

When the car is viewed from the rear, the narrow screen—again steeply raked—is shrouded by a sharp hood, and the tailgate's complex shutline snakes around the outer edge of the rear lights, which are themselves folded from the upright rear panel with a sharp crease around the continuation of the side feature line and on to the horizontal shoulder surface.

Lexus LX 570

Engine	5.7 V8
Power	286 kW (383 bhp)
Torque	544 Nm (401 lb ft)
Gearbox	6-speed automatic
Installation	Front-engined/four-wheel drive
Brakes front/rear	Discs/discs
Front tires	285/50R20
Rear tires	285/50R20
Length	4991 mm (196.5 in.)
Width	1971 mm (77.6 in.)
Height	1890 mm (74.4 in.)
Wheelbase	2850 mm (112.2 in.)

Despite its desire to project top-premium values of style, sophistication, and advanced engineering in the US luxury-car market, Lexus also needs to compete with the very large home-grown SUVs—typified by the Lincoln Navigator—which, until the recent crisis in confidence, continued to be a big profit center for North American producers.

The result is a third Lexus SUV line to slot in above the RX and GX series; the LX 570 is in effect an enhanced and much more luxurious version of the largest Toyota Land Cruiser, equipped with an even larger V8 engine displacing 5.7 liters. The Land Cruiser is an unashamedly bulky eight-seater with a simple body design that makes little concession to fashion or style. Lexus's treatment of the Toyota starting point is to soften its image by means of gentler, more rounded corners, and to add a front-end design unique to the upscale version. This sees a slightly different frontal style, with the bumper drawn further forward to allow a larger, rearward-sloping Lexus grille and neatly integrated lights alongside.

Broad running boards along the sills provide a handy step up to help ease access to the interior, even though the air suspension settles to its lowest setting when the vehicle stops.

The roof carries polished aluminum rails as opposed to the Toyota's black-finished items, and the interior, in genuine Lexus style, is truly vast and sumptuous.

Yet, in truth, this new version goes against the grain of Lexus ideology. Not only is it appearing at a time when even Americans are beginning to question the sense of vast gas-guzzling trucks, but also it incorporates little in the way of the increasingly successful design cues seen on other Lexus models. The designers may have managed to soften up the boxy shape somewhat, but it still looks heavy, upright and, ultimately, uninspiring.

Lincoln MKS

Design	Peter Horbury
Engine	3.7 V6
Power	201 kW (270 bhp) @ 6250 rpm
Torque	360 Nm (265 lb ft) @ 4250 rpm
Gearbox	6-speed automatic
Installation	Front-engined/all-wheel drive
Front suspension	MacPherson strut
Rear suspension	Multi-link
Brakes front/rear	Discs/discs
Front tires	255/45R19
Rear tires	255/45R19
Length	5184 mm (204.1 in.)
Width	1928 mm (75.9 in.)
Height	1565 mm (61.6 in.)
Wheelbase	2868 mm (112.9 in.)
Track front/rear	1651/1661 mm (65/65.4 in.)
Curb weight	1940 kg (4276 lb)

The MKS has been around in concept form since January 2006; now in its final production specification, it is luxury brand Lincoln's take on the standard Ford large sedan, the revived Taurus. To complicate the picture still further, there is a Mercury equivalent, too—the Sable. Those two models were themselves brought in to revive flagging sales of the Five Hundred and the Montego, when Ford dusted off comfortably familiar nameplates to help recognition by customers.

None of this model proliferation makes the designer's task any easier; in fact, in the case of the upmarket Lincoln brand there is the overriding need to differentiate the MKS clearly as a much classier and more luxurious product than the cheaper Ford and Mercury variants while at the same time respecting the fixed points of the platform and structure—and not adding too much extra cost.

It is a well-resolved basic body shape, if somewhat heavy at the rear. The Lincoln double-wing grille gives a distinctive aspect to the front, especially after the adoption of the central vertical body-color band separating the two wings. The Lincoln star emblem sits in the center. The 2006 concept had a full-width grille, which, while rather bolder, was less individual.

Gone, too, are the lateral air vents just behind the front wheel arches: now, a rather awkward-looking vertical Lincoln star emblem occupies that position. The chrome trim round the side windows widens toward the back, creating a hockey-stick effect and focusing attention on the rear compartment; the bland rear is marked by vertical taillights and a heavy chrome beam above the number plate, with the vertical Lincoln logo superimposed, again rather awkwardly.

Inside, the MKS is distinguished by nobler materials than in Ford versions, with reclaimed ebony and distinctively grained olive ash used on some grades. The driver benefits from Ford's Synch communications system, developed in conjunction with Microsoft, while a rearview camera mounted beneath the Lincoln star on the trunk lid aids reversing.

Lincoln MKT

Engine	3.5 V6
Power	309 kW (415 bhp) @ 5750 rpm
Torque	543 Nm (400 lb ft) @ 2000–5000 rpm
Gearbox	6-speed select-shift
Installation	Front-engined/all-wheel drive
Brakes front/rear	Discs/discs
Front tires	285/35ZR21
Rear tires	285/35ZR21
Length	5293 mm (208.4 in.)
Width	2000 mm (78.7 in.)
Height	1623 mm (63.9 in.)
Wheelbase	3066 mm (120.7 in.)
Track front/rear	1704/1702 mm (67.1/67 in.)

A Lear Jet for the road: that is how Ford-group design director Peter Horbury sees the latest concept from Lincoln, the US carmaker's domestic premium division. Translated into automobile terms, this becomes a luxury grand tourer, a new genre of premium vehicle that brings together elements from sports coupés, upscale limousines, and crossover utilities.

Not that there is anything remotely utilitarian about the svelte, bustle-backed body shape of this distinctive design—nor its ultrasumptuous interior, housing as it does four multiadjustable seats offering a standard of accommodation one might expect in a first-class airline cabin.

Crisp exterior detailing adds an impression of precision to the already striking design of the MKT. It is a shape that instantly demands a second and possibly a third glance: aside from the dramatic double-wing grille, with its graphic vertically finned teeth, the stepped waistline is an attention-getter, while the graceful bustle and full-width light band at the rear have an Italianate elegance that is rarely seen on American products.

Pronounced rear haunches and a strong, forward-leaning C-pillar give the design solidity and, when seen from the rear three-quarters, a coupé-like feel; one penalty of this is the large mass of sheet metal above the rear wheel, though this is relieved in some light conditions by the slight flare to the fender line.

Inside, the MKT mixes pale colors with smooth, sweeping forms and highlights in chrome, acrylic, and glass: the instrument cluster, says Lincoln, is inspired by Swarovski crystals. New, too, is the use of a Volvo-style floating center stack, allowing stowage space behind it.

The MKT is another Ford-group product to wave the environmental flag by ditching the thirsty V8 engine for the so-called EcoBoost V6. Offering 415 hp, the 3.5-liter powertrain nevertheless promises sparkling performance; but if this still troubles the green conscience of the privileged occupants aboard, they can always take comfort in the eco-friendly hand-knotted rugs made from banana silk that cover the floor.

Magna Steyr Mila Alpin

Design	Magna Steyr
Engine	3-cylinder 999 cc natural gas plus hybrid
Power	62 kW (84 bhp)
Gearbox	5-speed automated manual
Installation	Mid-engined/all-wheel drive
Length	3540 mm (139.4 in.)
Width	1703 mm (67 in.)
Height	1750 mm (68.9 in.)
Wheelbase	2225 mm (87.5 in.)
Track front/rear	1505/1505 mm (59.3/59.3 in.)
Curb weight	1086 kg (2394 lb)
0–100 km/h (62 mph)	11.5 sec
Top speed	140 km/h (87 mph)
Fuel consumption	5.4 l/100 km (43.6 mpg)
CO$_2$ emissions	99 g/km

Magna Steyr may not be a familiar name on Main Street or in car magazines, but it is an important player in the automotive industry. It is an expert in all-wheel-drive systems, supplies a wide range of components and systems to almost every automaker worldwide, and even builds complete cars—such as the Saab 9-3 Cabrio and the BMW X3—for outside customers.

The idea of the Mila Alpin concept is to demonstrate Magna's expertise in many of the fields in which it operates: four-wheel drive, alternative drive technologies, styling, and manufacturing. In effect, it is a rolling chassis into which various drive and energy-storage modules—liquid hydrogen, CNG (compressed natural gas), lithium-ion batteries, and gasoline—can be plugged.

The concept design is funky, purposeful, and fun, riding high on its chunky tires and clearly displaying its plug-in drive systems in a series of bays set in the sides of the black chassis section that forms the lower half of the vehicle. With good ground clearance, minimal overhangs, and a wheel at each corner, this has the configuration of a good climber and, mounted centrally, the three-cylinder engine enables a low center of gravity.

Less successful aesthetically is the white superstructure of the vehicle, which looks too toylike to match the tough technical strength of the lower half. The style is kept smooth and simple, with a large door at each side and a short hood ahead of the surprisingly upright windshield. Thick rear pillars house twin taillights at each side and contain air scoops to feed fresh air to the cabin as well as to the engine, allowing the machine to wade to a depth of 20 inches. At the rear the exhaust is neatly recessed into the angled valance so as to avoid damage on rough terrain.

Overall, this is the impressive first draft of a concept that could yield a very attractive, versatile, and useful light mountain-climber promising real ability.

Mazda6

Design	Youichi Sato
Engine	2.5 in-line 4 (1.8 and 2.0 also offered)
Power	125 kW (168 bhp) @ 6000 rpm
Torque	226 Nm (167 lb ft) @ 4000 rpm
Gearbox	6-speed manual
Installation	Front-engined/front-wheel drive
Front suspension	Double wishbone
Rear suspension	Multi-link
Brakes front/rear	Discs/discs
Front tires	195/65R16
Rear tires	195/65R16
Length	4735 mm (186.4 in.)
Width	1795 mm (70.7 in.)
Height	1440 mm (56.7 in.)
Wheelbase	2725 mm (107.3 in.)
Track front/rear	1560/1560 mm (61.4/61.4 in.)
Top speed	220 km/h (137 mph)
Fuel consumption	8.1 l/100 km (29 mpg)
CO_2 emissions	192 g/km

The outgoing Mazda6 was the model that revived Mazda's fortunes after a long sales slide in Europe and globally: it was neat, good-looking, and well made, drove well, and had lasting style. So it is little surprise that for the second generation of this cornerstone upper-medium-sized model Mazda has stayed close to the original in terms of shape, stance, and detailing.

Important, however, is the fact that while the new car is clearly recognizable as a Mazda6, it does move the design on in several key respects. In particular, the design has clearly grown up: there is an air of Lexus-like maturity, especially from the rear. Chief designer Youichi Sato describes the design theme for the new Mazda6 as "bold and exquisite." The latter quality is perhaps easier to spot; there is a definite dynamic poise to the car, something Mazda's key designs have often succeeded in achieving, and sharp-edged lines and decorative elements give added energy to the shape.

Mazda was one of the first to introduce the new front fender style on its 2 hatchback: the Mazda6 takes this a step further, with the arched style becoming one of the car's dominant features. The rear of the semicircular crease formed by the arch then runs straight backward as a feature line, taking in the door handles and meeting the forward point of the rear lights. A further classy detail is seen on the outboard lower air inlets at the front, where a neat chrome strip divides the vents, pointing toward a square fog light at the inner edge.

The station-wagon version is equally attractive, with a rising window line and a stylish V-shape to the base of the tailgate window. The interior design, with some use of bright metal highlights, is clean and tidy though lacking in genuine distinction.

Mazda Furai

Design	Franz von Holzhausen
Engine	Three-rotor rotary
Power	336 kW (450 bhp)
Installation	Mid-engined/rear-wheel drive
Front suspension	Double wishbone
Rear suspension	Double wishbone
Brakes front/rear	Discs/discs
Width	2000 mm (78.7 in.)
Height	1000 mm (39.4 in.)
Top speed	290 km/h (180 mph)

From the point of view of design progression, the Furai is perhaps the least immediately impressive of Mazda's recent run of its "Nagare," or wind-inspired, concept vehicles. Its fundamental proportions are familiar from the world of endurance-racing sports cars; indeed, it is directly built on the real-life carbon-composite tub of a Courage C65 racer, with the only major changes being those designed to enhance the aerodynamics and, in line with the Nagare philosophy, give visual representation to the flow of air.

On closer examination, the Furai does in fact reveal a wealth of fascinating detail, such as the complex interplay of the flowing forms along the sides, the intriguing tiny rearview mirrors formed out of the cant rails, and the beefy and businesslike rear end dominated by big venturi tunnel exits and the massive, overwidth spoiler wing.

The front is particularly graceful, with its pronounced point, which carries the Mazda logo, and the flowing, swirling leaves that sweep out sideways and crisscross to break up the expanse of the lateral air intakes and form the shape of the headlights. The shelflike front splitter projects out from under the nose, just as on a racing car—an arrangement entirely unsuitable for a road vehicle.

The Furai's sides again portray the flow of air, a series of feature lines developing just behind the front wheel to crystalize into horizontal vanes sweeping into the large, gashlike air intakes just ahead of the rear wheels: the profile of these vanes is picked out with long lines of LEDs.

Access to the very tight two-seater cabin is through a pair of doors that hinge upward from the rear of the windshield. The cockpit is cramped and uncompromising, as one would expect on a racing car, but is clad in plush red suede. The driver sits a long way forward, the triple-rotor race engine occupying a significant portion of the car's length.

Mazda Taiki

Design	Yamada Atsuhiko
Engine	RENESIS rotary engine
Gearbox	7-speed manual
Installation	Front-engined/rear-wheel drive
Front suspension	Double wishbone
Rear suspension	Double wishbone
Brakes front/rear	Discs/discs
Front tires	195/40R22
Rear tires	195/40R22
Length	4620 mm (181.9 in.)
Width	1950 mm (76.8 in.)
Height	1240 mm (48.8 in.)
Wheelbase	3000 mm (118.1 in.)

For the fourth in its spectacular series of Nagare design studies aiming to express the flow of air as a design theme, Mazda has taken the concept to its most extreme expression so far. The Taiki, shown at Tokyo in 2007, is a vision of a sports car for the sustainable society of the future and is appropriately dramatic in its configuration as well as its execution.

This is a vehicle that, at first glance, looks more like a fighter plane than a car. A wide grille mouth, with elegantly curved crossbars that resemble a bird's wings flapping in flight, leads to the short hood and the wide-set front wheels, which have spokes resembling the fan blades in a jet engine rotor; however, everything aft of the front wheels is different.

A narrow, low-set, all-glass bubble canopy houses the two occupants, setting the tone for the rest of the planelike fuselage, which tapers in tightly to reach a point behind the rear axle. The rear wheels are carried on short winglike extensions from the waistline of the fuselage and are completely shrouded by smooth fairings, much like the undercarriage of some light aircraft: the outer edge of the tire is all that can be seen. No suggestion is given of how rear suspension movement can be accommodated.

At the rear, the view is more astonishing still. There is very little body mass apparent, and all that meets the eye is the sinuous M-shape of the narrow strip of taillight that follows the rear contours of the trailing edge of the rear wheel spats, rising upward into the fenders and thence curving rearward to outline the pointed tail section. Just visible in the air tunnels between the wheels and the fuselage are the exhaust outlets and the suspension elements.

Powered by Mazda's next-generation rotary engine, front-mounted but driving the rear wheels, this design may be impractical and fantastical, but it is undeniably futuristic.

Mercedes-Benz F700

Every so often, a design comes along that seems like a step into the future. The Mercedes F700 is just such a design. Here's why. First, it has an innovative powerplant that combines gas-engine cleanliness with a diesel's economy and flexibility; secondly, this is all wrapped in a body that, while perhaps not conventionally pretty, clearly places a higher priority on resource conservation and aerodynamic efficiency than on ostentation or luxury-class gravitas. Conceivably, this could be an early blueprint for the next-generation S-Class.

The most striking aspect of the body design is the way it is made up of gentle organic forms: everything is soft, smooth, and flowing, rather like a shoal of fish in the ocean. There's scarcely a straight line to be seen; even the waistline and the cant rail are curved, while the tail of the car curves away in a manner rarely seen among status-conscious limousines.

The effect is unusual, to say the least: the whole body appears to be sagging over the wheels, as if it is shrugging its shoulders to release any tension from within. For the historically minded, it has echoes of Pininfarina's controversial "banana" car of the 1970s.

The only design cue anchoring this flowing shape as a Mercedes is a very large grille, plunging almost to ground level; large headlights sit on the shoulders that form the wheel arches, again separating the elements that make up the body, while the body-color band along the length of the rear lights echoes that on the headlights.

The doors and the seating arrangement are asymmetric, with the right-hand side having a rear-hinged rear door and two passengers facing one another, while the left is more conventional. For a car that has a multitude of complex systems, including one that can read the road ahead to prepare the suspension for approaching bumps, the F700's interior is remarkably calm and simple—a good omen for the future, no doubt.

Design	Peter Pfeiffer
Engine	1.8 in-line 4 Diesotto
Power	178 kW (238 bhp)
Torque	400 Nm (295 lb ft)
Brakes front/rear	Discs/discs
Front tires	195/55R21
Rear tires	195/55R21
Length	5180 mm (203.9 in.)
Width	1960 mm (77.2 in.)
Height	1438 mm (56.6 in.)
Wheelbase	3450 mm (135.8 in.)
Curb weight	1700 kg (3748 lb)
0–100 km/h (62 mph)	7.5 sec
Top speed	200 km/h (125 mph)
Fuel consumption	5.3 l/100 km (44.4 mpg)
CO_2 emissions	127 g/km

Mercedes-Benz Vision GLK Freeside

The long and cumbersome title is Mercedes' way of stating that this concept is a near-production preview of its belated competitor to the BMW X3 and Land Rover Freelander. It is also the basis from which Smart was to have developed its ForMore leisure SUV.

Those watching the debut of the new model at the Detroit show in January 2008 were somewhat surprised at what was unveiled. Rather than the smooth, family-friendly, non-confrontational crossover SUV that most had been expecting, what appeared was a much more traditional off-roader, complete with rugged, square-themed styling, an upright rectangular glasshouse, and heavy 4x4 design cues, such as blistered wheel arches and raised ground clearance. The big three-bar grille, too, appeared to be lifted straight from Mercedes' classic Jeep-like G-Wagen series.

Under the skin is a catalog of advanced hardware, running to a 2.2-liter Bluetec diesel engine, seven-speed transmission, the clever 4Matic four-wheel-drive system, and special suspension and steering systems.

Mercedes' line on the exterior style of the GLK is that its echoing of the angular look of the style-icon G Series will spell out the model's strength and character, with such features as the upright front and short front and rear overhangs further emphasizing its true off-road ability. Certainly, the interplay of the many conflicting taut lines and right angles makes for an unusual, if somewhat old-fashioned, look; the nose and hood are particularly heavy and chunky, and far removed from the more delicate, sympathetic designs rival brands are now trying to move toward. The interior again has more than its fair share of straight lines and edges, especially on the instrument panel and center display hood.

Judging by this near-final concept, Mercedes' intention must be to differentiate the production GLK from its softer-edged competitors by reinforcing the messages of ruggedness and off-road ability, rather than those of comfort and social compatibility.

Design	Peter Pfeiffer
Engine	2.2 in-line 4 diesel
Power	127 kW (170 bhp)
Gearbox	7-speed automatic
Installation	Front-engined/all-wheel drive
Brakes front/rear	Discs/discs
Front tires	20 in.
Rear tires	20 in.
Length	4520 mm (177.9 in.)

Mini Clubman

Engine	1.6 in-line 4
Power	130 kW (175 bhp) @ 5500 rpm
Torque	240 Nm (177 lb ft) @ 1600–5000 rpm
Gearbox	6-speed manual
Installation	Front-engined/front-wheel drive
Front suspension	MacPherson strut
Rear suspension	Longitudinal arms
Brakes front/rear	Discs/discs
Front tires	195/55R16
Rear tires	195/55R16
Length	3958 mm (155.8 in.)
Width	1683 mm (66.3 in.)
Height	1432 mm (56.4 in.)
Wheelbase	2547 mm (100.3 in.)
Track front/rear	1453/1461 mm (57.2/57.5 in.)
Curb weight	1280 kg (2822 lb)
0–100 km/h (62 mph)	7.6 sec
Top speed	224 km/h (139 mph)
Fuel consumption	6.3 l/100 km (37.3 mpg)
CO_2 emissions	150 g/km

Few recent designs can have been less of a surprise than the Mini Clubman. Born out of a desire to stretch the Mini range in a more family-oriented direction, the Clubman was effectively previewed by the series of concept models exhibited at international motor shows in 2006 and 2007.

The Clubman is a station wagon but, as with everything Mini, it represents a very different take on the idea. Inspired by the original Mini Traveller of the 1960s, the Clubman—another recycled Mini title—has a longer wheelbase for more rear-seat room, as well as extra overhang for respectable space in the cargo bay. But rather than follow the herd and go for a sporty lifestyle-wagon-type angled tailgate, Mini looked to its own heritage, chopping the tail off vertically and inserting a pair of side-hinged doors—just as on the original.

Not only that, but there's a novel side-door solution, too, at least on the right-hand side. Here, Mini has added a small extra "club" door, hinged at the back, which can be opened only once the main door is open. This noticeably improves access to the rear—important for a parent securing the seatbelt round a child in the back seat.

In the Clubman, Mini has invested heavily in solving problems that people did not realize existed. And while those solutions cannot make the Mini as practical as a mainstream hatchback, they are highly original and a surefire talking point. They are thorough and assuredly expensive, too; the cargo doors, for example, are elaborately hinged round the taillights so the lights remain visible even when the doors are open. And then there's the issue of the lack of the small extra door on the left: moving the fuel filler and perhaps even the tank would have been too costly, even for BMW's Mini engineers.

Mitsubishi Concept-cX

Engine	1.8 in-line 4 diesel
Power	101 kW (136 bhp) @ 4000 rpm
Torque	280 Nm (206 lb ft) @ 2000 rpm
Gearbox	6-speed dual clutch
Installation	Front-engined/all-wheel drive
Front tires	225/45R19
Rear tires	225/45R19
Length	4100 mm (161.4 in.)
Width	1750 mm (68.9 in.)
Height	1550 mm (61 in.)
Wheelbase	2525 mm (99.4 in.)
Track front/rear	1510/1510 mm (59.4/59.4 in.)
Curb weight	1360 kg (2998 lb)

This striking-looking family-sized car is Mitsubishi's idea of a new market opportunity—a sports hatchback crossover. Translated into everyday language, this could be a hot-hatch with a hint of off-road ability, or an SUV very strongly biased toward on-road dynamics. Nissan is already enjoying big success with its Qashqai (*The Car Design Yearbook 6*), which adds an adventurous off-road flavor to a family hatchback, while Toyota and Land Rover will launch compact sports SUVs with a clear on-road bias.

The Concept-cX moves the game along from the Shogun Pinin, Mitsubishi's original venture into the compact SUV subsegment. Whereas that model was clearly SUV-derived, the cX has proportions much more like a normal family hatchback. The only real hints at any off-road capability are the big wheels and tires, the slightly raised ground clearance, and the polished metal plates that protect the sills and the front and rear aprons.

Dominating the car's appearance is the very large square grille, dark and void of any decoration; aggressive headlamps peer from under the corners of the clamshell hood, and the wide front wheel arch flows back into the body side to form a long and distinctive feature line. The glasshouse is very much that of a small hatchback, with blacked-out B- and C-pillars to create continuity. The rear, however, is very unusual: the black effect is carried the full depth of the tailgate, down to bumper level. This, along with the separate look of the rear fenders, gives a retro feel to the rear.

The interior is unusual, too, dominated by tan-colored upholstery (derived from plant-based fibers) and by the large upright oval housing in the center of the dashboard that carries the minor controls and displays. A further novelty is the return to the idea of a full-width bench seat in the front: the gear lever and small switches are housed in a raised central portion.

Mitsubishi Concept RA

Design	Gary Ragle
Engine	2.2 in-line 4 diesel
Power	150 kW (201 bhp)
Torque	420 Nm (310 lb ft)
Installation	Front-engined/four-wheel drive
Brakes front/rear	Discs/discs
Front tires	285/30R21
Rear tires	285/30R21
Length	4445 mm (175 in.)
Width	1895 mm (74.6 in.)
Height	1315 mm (51.8 in.)
Wheelbase	2635 mm (103.7 in.)
Track front/rear	1605/1605 mm (63.2/63.2 in.)

Following on from a series of concept studies exploring themes in the areas of family sedans and compact SUVs, Mitsubishi has returned with the Concept RA to a genre in which it has not been present for many years: that of the pure sports car.

Yet, as is so often the case with Mitsubishi, there is an unusual twist on the theme. The RA, bravely, sets out to show that thrilling performance and environmental responsibility need not be mutually exclusive concepts. The former is assured by a powerful engine and a whole raft of advanced chassis technologies drawn from the iconic, rally-inspired Evo-X model, while the RA's green credentials are guaranteed by a 2.2-liter clean diesel engine—unusual in a sports car—and a twin-clutch automated manual transmission.

The teasing is taken a step further in the RA's strongly cab-forward side profile. With equally balanced visual masses at front and rear of the cabin, itself a simple sweeping arc linking A-pillar, cant rail, and B-pillar, the observer is left wondering whether the design is front- or mid-engined. A shift round to the front soon provides the answer: the matt black hood contains a large rectangular cutout in the center, through which projects the smoothly styled top end of the diesel engine. Also black, the engine cover integrates well, though it could never pass pedestrian safety tests in Europe.

Visually dominant in the design are the large 21-inch wheels and their prominent arches, which bulge horizontally out from the car's curving sides as if to emphasize the extreme width of the tires. The rest of the bodywork wraps tightly around the essential mechanical elements to create a compact-looking sports car clearly influenced by the Lancia Stratos. There is also a more contemporary similarity to Peugeot's 308 RC Z concept, also front-engined, of autumn 2007, especially in the rear view, where the long rear deck and extreme slope of the rear window are distinguishing features.

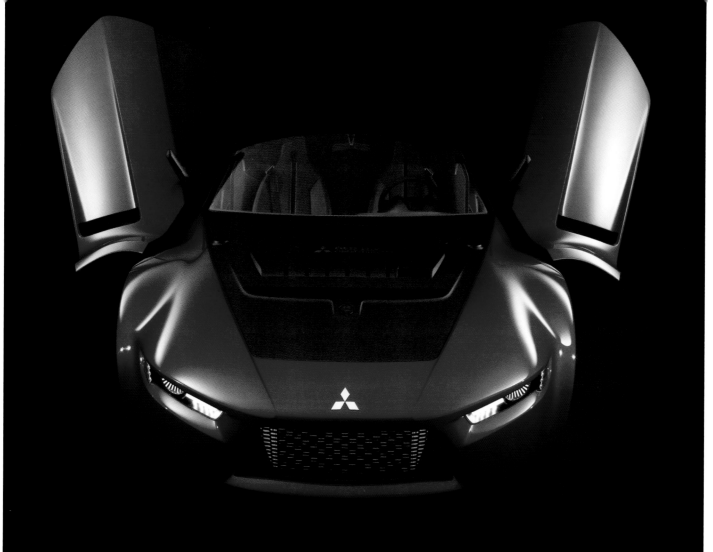

Mitsubishi Concept ZT

Engine	2.2 in-line 4 diesel
Power	140 kW (188 bhp)
Torque	400 Nm (295 lb ft)
Installation	Front-engined/all-wheel drive
Brakes front/rear	Discs/discs
Front tires	255/35R20
Rear tires	255/35R20
Length	4950 mm (194.9 in.)
Width	1820 mm (71.7 in.)
Height	1440 mm (56.7 in.)
Wheelbase	2815 mm (110.8 in.)
Track front/rear	1555/1555 mm (61.2/61.2 in.)

Although it shares its name with that of an obsolete MG sedan model, there is much more Audi and Volvo than MG in the look of the Concept ZT, Mitsubishi's proposal for a large sedan to edge its offerings back up toward the near-premium market.

The ZT possesses a calm and an elegance that are absent from Mitsubishi's more aggressive smaller products. The lines and the proportions balance in a dignified, tranquil way, with a smoothly arched DLO profile that echoes the Audi A6, the A-pillars swept a long way rearward to generate an impression of luxury. The only hints of pushiness are the thrust-forward grille—similar in shape to the new Lancer family but more subtly drawn—and the broody eyebrows over the headlights.

Yet the very long overhangs at front and rear give a clue that this is a big car spun off the architecture of a smaller one; the fact that the rear cabin and the rear door shutlines extend back to the rear wheel center line is further evidence. Such visual devices as the clear, undercut feature line flowing rearward from the front wheel arch serve to stretch the look of the car; additionally, a chrome strip along the sills is taken forward on the side of the front apron, where it turns through a rather contrived zigzag across the side air inlets to meet a higher crossbar in the lower central grille.

If the exterior of the ZT is conservative and restrained, its interior is just the opposite. A lurid wall-to-wall electroluminescent dashboard greets the driver, its myriad displays, gauges, and readouts sparkling like Times Square's advertising billboards at night; as if that were not enough, the blue neon theme is continued into the center-stack input keys, the gear selector controls, the window controls alongside the driver, and the luminous blue piping that bridges the dashboard and doors at shoulder level.

Nissan Forum

Gearbox	CVT
Installation	Front-engined/front-wheel drive
Brakes front/rear	Discs/discs
Front tires	245/45R20
Rear tires	245/45R20
Length	4990 mm (196.4 in.)
Width	2115 mm (83.3 in.)
Height	1763 mm (69.4 in.)
Wheelbase	3075 mm (121.1 in.)

One of the most striking concepts at the 2008 Detroit show, the Nissan Forum shows the great lengths US designers are going to in their struggle to get away from the dull, boxy, "soccer mom" stereotype that blights the appeal of such conventional large minivans as the Chrysler Voyager.

The Forum could hardly be more different. From the big, black grin of the wraparound grille, the lines flow backward in a series of soft, fluid forms; the surface language is both attractive and intriguing, with gently winding surfaces giving movement to the vehicle's large flanks. Black upper pillars help to make the car look more streamlined by taking out the vertical visual elements, and the dynamism continues at the rear, where the beltline drops to provide a deeper, near-vertical tailgate glass that is reminiscent of a Renault Mégane.

The front is a shade more aggressive. The broad grille, with the triangular headlight units set at its upper corners, plunges downward into the front valance, itself marked by deep, wide-set triangular air intakes and lights; above the grille the complex hood is crowned in the middle and also rises at the edges to form the A-pillars.

A single long handle opens the side doors, with the rear door sliding back tracklessly and giving unimpeded access to a big, sumptuous interior, which houses six luxurious seats, each fitted with its own built-in six-way Bose speaker system.

The instrument panel again uses fluid forms, but is perhaps less aesthetically successful than the exterior. Rich cream and wood colors combine with tanned leather and the black dash capping to create a premium ambience, yet the combination of the strong vertical wood grain with the horizontal shapes and the floating instrument pack results in a clash of messages. Fittingly for a vehicle designed by Nissan's California studio, there is even a microwave oven in the center console.

Nissan GT-R

Design	Hiroshi Hasegawa
Engine	3.8 V6
Power	353 kW (473 bhp) @ 6400 rpm
Torque	588 Nm (433 lb ft) @ 3200–5200 rpm
Gearbox	6-speed manual
Installation	Front-engined/four-wheel drive
Front suspension	Double wishbone
Rear suspension	Multi-link
Brakes front/rear	Discs/discs
Front tires	255/40ZR20
Rear tires	285/35ZR20
Length	4655 mm (183.3 in.)
Width	1895 mm (74.6 in.)
Height	1370 mm (53.9 in.)
Wheelbase	2780 mm (109.4 in.)
Track front/rear	1590/1600 mm (62.6/63 in.)
Curb weight	1740 kg (3836 lb)

Though the Nissan GT-R has never before been made in left-hand drive, successive generations of this iconic Japanese sports coupé have helped to build up a global cult following, and a fanatical one at that. With the previous GT-R having been withdrawn several years ago, its replacement has been awaited with growing impatience by this global fan base, especially as Nissan allowed it to be known that the new car was being targeted at no less a rival than the Porsche 911 Turbo.

Performance-minded fans will certainly not be disappointed with the design that was unveiled with a big fanfare at the 2007 Tokyo show: it matches the Porsche horsepower for horsepower and turbo for turbo, and it has the measure of the 911 in terms of speed and acceleration.

But where the two are at opposite ends of the spectrum is in the area of style. The Nissan is, frankly, brutal, its exterior is not refined and sophisticated but functional and aerodynamic. The design is built up of chiseled-off polygon shapes, giving a blocky effect, and large vents at the back of the front fenders highlight the amount of cooling air that is needed to pass through the potent engine and hefty brakes. Everywhere you look, there are vents, slats, and add-ons, though there is a smooth shape underneath, and the GT-R hallmark of quadruple round rear lights is faithfully incorporated.

The rear aspect is particularly menacing, with four large exhausts jutting out through the blackened apron at either side of the central diffuser, and a full-width, purposeful spoiler atop the trunk lid. The window line begins to drop immediately behind the A-pillar, concentrating the visual focus on the front seats.

The few isolated red highlights aside, the interior is all gray, with a clear focus on the needs of the enthusiast driver: the fat-rimmed three-spoke steering wheel carries numerous minor control functions, while the all-important rev counter is set at dead center in the driver's line of sight.

Nissan Maxima

Design	Giovanni Arrobia
Engine	3.5 V6
Power	216 kW (290 bhp)
Torque	354 Nm (261 lb ft)
Gearbox	CVT
Installation	Front-engined/front-wheel drive
Front suspension	MacPherson strut
Rear suspension	Multi-link
Brakes front/rear	Discs/discs
Front tires	245/40R19
Rear tires	245/40R19
Width	1859 mm (73.2 in.)
Wheelbase	2776 mm (109.3 in.)
Track front/rear	1585/1585 mm (62.4/62.4 in.)
Fuel consumption	10.2 l/100 km (23.1 mpg)

The Nissan Maxima was sold for a while in Europe, never with conspicuous success, but it has been a staple of the North American market for as long as anyone can remember. Over its seven generations in the USA it has changed its image almost as many times, switching identities regularly between its original incarnation as a sports sedan and a more conventional mainstream sedan.

The seventh generation sees the big Nissan embrace the allure of a sleek, coupé-like luxury sedan once more, and it comes across as lower, sleeker, and sportier than many of its often overweight competitors. In the new design the front, especially the central grille, is similar to that of a Volvo S80, the thrust-forward rectangular air intake leading into a broad, gently contoured hood. The headlights, set back rather like the Volvo's, are nevertheless very different: they slash back in a boomerang shape into the front fenders in a way that is unique to Nissan.

As with many other current designs, the sides of the Maxima show the influence of BMW, the crisp feature line running rearward from the indicator lens fading out around the B-pillar, only to reappear in the rear door to provide a strong shoulder over the rear wheel arch. A recess at the base of the doors allows light effects to play on the lower sides.

Rear lights somewhat in the style of the Mercedes S-Class cut into the rear fenders and slant forward, while on their inner edges they frame the recessed trunk lid. The rear deck rises higher than the rear fenders to meet the back window and C-pillars; this is often a tricky area to handle, and Nissan's solution is no more successful than those of other carmakers.

The interior, surprisingly for a big sedan, forsakes grafted-on wood, chrome, and aluminum panels, instead giving a more sporty—though hardly classy—impression through the use of gray and black.

Nissan Mixim

Japan's Nissan has been making some pretty ambitious promises about moving into the production of electric cars; it is this commitment—allied to the realization that in Japan and some other mature markets young people are no longer excited by cars—that has led the company's designers to come up with the Mixim concept.

The role of the Mixim, says Nissan, is to inspire young people to put their computers to one side and re-engage with the automobile. As such, it uses the most up-to-date electric technology, with a lithium-ion battery powering motors on both axles, giving the effect of four-wheel drive.

In its determination to be novel, Nissan has given the Mixim a three-seater layout like that of the McLaren F1. The driver sits centrally, holding the twin handgrips and with a good view of the wraparound instrument pod; the two passengers sit beside and slightly behind, while a fourth occasional seat is located directly behind the driver.

The diamond shape of the seating pattern is a theme that dominates the Mixim's design both inside and out. Almost every surface or detail is diamond-shaped or triangular: even the overall coupé profile, characterized by a dramatic thick cant rail that cuts diagonally downward and backward from the A-pillar, is triangular in character. Spiky triangular windows above the cant rail bring more light into the cabin, while the large doors hinge upward and forward, again creating a dramatic shape.

Front lighting is handled by six diamond-shaped and two long triangular headlights, echoed in triangular recesses in the body side at the trailing edge of the door. Solid disc wheels bring to mind the rotors of certain electric motors.

Overall, the most successful area of the Mixim is the rear, where more curved surfaces are evident and where the crown of the roof gives the wide rear window a pleasing aspect.

Design	Shiro Nakamura and François Bancon
Engine	Electric, lithium-ion battery
Front tires	18 in.
Rear tires	19 in.
Length	3700 mm (145.7 in.)
Width	1800 mm (70.9 in.)
Height	1400 mm (55.1 in.)
Wheelbase	2530 mm (99.6 in.)
Curb weight	950 kg (2094 lb)

Nissan Murano

It is to Nissan's credit that the original Murano, launched to some controversy in 2004, still looks fresh and unusual: it was one of the first of the so-called crossover designs to use a lighter and more refined car platform rather than a crude truck-type chassis for a family-oriented SUV. Only the lack of a diesel-engine option held it back in Europe, while in North America it was a big success.

The made-over 2009 model year Murano retains the distinctive Murano flavor, with similar proportions and the same air of sophistication rather than rugged functionality. It no longer looks like such a standout design, but that is probably down to the fact that competitors have advanced quite significantly in the intervening years; in isolation, it is still a well-turned-out and attractive vehicle.

The new face of the Murano intertwines the grille and the headlamps to create a subtly different look; the first-generation model had headlamps that extended back up into the wheel arch, while this version allows the lamps to dip into the bumper area. Strong vertical chrome slats characterize the new grille—a contrast to the mesh effect of the old car.

The surfacing around the front wheel has been simplified to a single feature line circumscribing the whole arch; the surface created below this line fades out as it flows into the door, only to reappear around the rear wheel arch, too.

Though the silhouette of the passenger compartment is carried over, the rear treatment is again subtly different: the lamps and the screen are the most noticeable change, with the tailgate shutline now running through the lamps and the screen swapping its simple curved base for a faceted design with an odd-looking arc pushing upward into its base section.

Inside, the design is effective enough but lacking in true flair; however, Nissan's angled multifunction controller panel continues to be one of the industry's better examples.

Engine	3.5 V6
Power	198 kW (265 bhp)
Torque	337 Nm (248 lb ft)
Gearbox	CVT
Installation	Front-engined/all-wheel drive
Front suspension	MacPherson strut
Rear suspension	Multi-link
Brakes front/rear	Discs/discs
Front tires	P235/55R20
Rear tires	P235/55R20
Length	4788 mm (188.5 in.)
Width	1882 mm (74.1 in.)
Height	1699 mm (66.9 in.)
Wheelbase	2825 mm (111.2 in.)

Nissan NV200

The idea behind this show concept, says Nissan, is to propose a new type of van that will form the basis for a "smart business tool" for a new generation of active professionals. For the NV, Nissan decided to outfit the van to suit the needs of a professional ocean photographer, though, mercifully, the company stopped short of trying to make it amphibious or able to dive to the depths for striking authentic underwater shots.

The concept is of a mobile office and van in one. There is a clear separation between the front of the vehicle—the mobility part—and the rear portion, used for professional activities. Taking the photographer as an example, his vehicle needs to carry such equipment as underwater cameras, lights, scuba-diving gear, an underwater scooter for reconnaissance trips, computers, a change of clothes, and food and water. For longer assignments, somewhere to sleep is also important.

All this is accommodated in a clever pod that pulls out rearward from the tailgate, supported on an internal rail and legs that brace it to the ground. This gives easy access from ground level to all the equipment, and leaves the interior clear for desk work.

Rugged bumpers and wheels (with six arms that appear to grip the tire itself, octopus-fashion) and roof-mounted expedition lighting all add to the sense of adventure, while when closed the rear has a vaultlike look to it, emphasizing the important content within.

There is much ingenuity in the design and planning of this concept—something that could be turned to good use in a range of retail, trade, and adventure applications.

Design	Ryoichi Kuraoka and Stephane Schwarz
Length	4380 mm (172.4 in.)
Width	1700 mm (66.9 in.)
Height	1840 mm (72.4 in.)
Wheelbase	2800 mm (110.2 in.)

Nissan Pivo 2

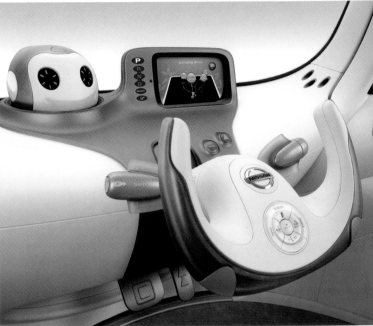

Design	Masato Inoue
Engine	Electric
Installation	Individual electric motors in each wheel
Length	2600 mm (102.4 in.)
Width	1690–2200 mm (66.5–86.6 in.)
Height	1700 mm (66.9 in.)
Wheelbase	2000 mm (78.7 in.)

Nissan's original Pivo was a huge hit when demonstrated at the 2005 Tokyo show. What truly captured the public imagination was how easy the tiny electric car promised to be in tight urban conditions. Most remarkably, the Pivo offered a tempting end to the tiresome chore of reversing: instead of craning his or her neck to back out of a parking space, the driver just rotated the globelike passenger compartment until it was facing the other direction, and drove off. No reverse gear was needed: the chassis automatically knew the direction the cabin was facing, and operated accordingly.

The Pivo 2 goes one better. The cabin still rotates, but the wheels now swivel, too: this enables the vehicle to move sideways, like a crab, for even more remarkable maneuverability. Each of the four wheels has its own electric motor, equalizing the drive forces, and stability is enhanced under acceleration, braking, and cornering by the appropriate wheels moving outward from the chassis.

Inside the bubble of the three-seat cabin, the driver sits centrally and forward of his or her companions: the style is clean, cozy and snug in cream and pale blue, with the orange controls adding a playful element reminiscent of children's games. The interior has a further, intriguing innovation: a monkey-like "Robotic Agent" that peers out of the window and whose role is to make the driver's journey less stressful. This involves giving navigation instructions, finding parking spaces, and even engaging the driver in conversation—in either English or Japanese. Nissan claims, too, that the Agent is able to gauge the driver's mood through facial expressions, and react accordingly.

This is a concept that builds on the originality of the first version, and would be entirely practical in a microenvironment where there was no risk of being hit by a large truck.

Nissan Round Box

Design	Kaoru Sato
Gearbox	CVT
Length	3650 mm (143.7 in.)
Width	1750 mm (68.9 in.)
Height	1530 mm (60.2 in.)
Wheelbase	2430 mm (95.7 in.)

Firmly in the category of strange Tokyo-specific concept cars is this creation from Nissan, the design aim of which is to serve as a "youth-oriented compact convertible to provide occupants with a sensation of speed and exhilaration."

This might sound like the recipe for a high-powered sports roadster, but Nissan begs to differ. Instead, the Round Box is a curious amalgam of styles, as if an expanded Smart two-seater had been grafted onto a sports car chassis—and the roof had been removed at the same time. The result is an eye-catchingly odd mélange of ingredients that is somewhere between brilliant and borderline crazy, though no one could describe it as elegant or attractive.

The two main elements of the body are the broad lower section, carrying the grumpy-looking frontal grille and the wheels, and the narrower, more upright cabin module. The two-layer effect is clearest at the sides, where there is a step between upper and lower sections; at the front and rear, the hood and trunk fairings perform the tricky task of linking the two structural shapes. Set into the doors at foot level are rectangular windows, to allow occupants to experience a sense of the road rushing past at speed.

The solid base section has a skateboard-like gravity to it, with the softer, playful, and more dynamic cabin mounted above; from the rear, eyes are drawn to the white surround encircling the rear screen; this in fact houses the rear lights. White detailing is carried across to the door handles and inside the cabin, where bold white-and-orange patterns are the theme of the five seats.

The controls offer a further nod to the youth vote. The steering element—it is certainly not a wheel—resembles a sophisticated computer-gaming handset, while the all-electric dashboard incorporates an interactive game to be played between the front passenger and those in the rear.

Opel/Vauxhall Agila

Design	Belinda Müller
Engine	1.2 in-line 4 (1.0 gasoline and 1.3 diesel also offered)
Power	63 kW (85 bhp) @ 6000 rpm
Torque	114 Nm (84 lb ft) @ 4000 rpm
Gearbox	5-speed manual
Installation	Front-engined/front-wheel drive
Front suspension	MacPherson strut
Rear suspension	Torsion beam
Brakes front/rear	Discs/discs
Length	3740 mm (147.2 in.)
Width	1680 mm (66.1 in.)
Height	1590 mm (62.6 in.)
Wheelbase	2350 mm (92.5 in.)
Track front/rear	1470/1480 mm (57.9/58.3 in.)
0–100 km/h (62 mph)	12 sec
Top speed	174 km/h (108 mph)

Opel and Vauxhall in Europe have relied on Japan's Suzuki for their entry-level models for some years, but the new Agila is a far cry from the narrow, boxy, and unloved Agila of old.

That model, which was based on the cheap and not very cheerful Wagon R+, was appreciated mainly by older buyers with no interest in aesthetics or dynamics, let alone brand image or style. The new edition shoots to the opposite end of the age range and takes as its inspiration the new and youthful Suzuki Splash, previewed as a concept in 2006 and launched into production in late 2007. Differences between the two cars, which are produced at the same factory in Hungary, are confined to the "soft" areas of the nose and tail; structurally they are identical.

To say that the new Agila is more dynamic-looking than the vehicle it replaces is an understatement: the new version stands tall and proud, with an eager, bold look thanks to its almond-shaped "eyes," its minimal rear overhang, and its sportily wide-set back wheels. The tall rear lights resemble arrows pointing forward and down along the side feature line to the front wheels; the large door surfaces are broken up by a sculpted form that alters the way the light is reflected, while the rear door openings are cut into the roof to ease access. The angles of the door shutlines, the rear quarter-light, and the rear light give the car a quirky leaning-back stance, accentuating the roomy look.

Sadly, this quirkiness is not followed through into the design of the interior. Apart from the bright colors (on some versions) and the odd placement of the rev counter on the top of the dashboard, the Agila's cabin is short of originality as far as its design is concerned. This could prove a drawback in its target market of young women.

Opel/Vauxhall Flextreme

Design	Anthony Lo
Engine	1.3 in-line 4 diesel hybrid with lithium-ion battery
Power	136 kW (182 bhp)
Torque	370 Nm (273 lb ft)
Gearbox	Front engine driving generator powering electric wheel motors
Front suspension	MacPherson strut
Rear suspension	Torsion beam
Brakes front/rear	Discs/discs
Front tires	195/45R21
Rear tires	195/45R21
Length	4555 mm (179.3 in.)
Width	1836 mm (72.3 in.)
Height	1487 mm (58.5 in.)
Wheelbase	2725 mm (107.3 in.)
0–100 km/h (62 mph)	9.5 sec
Top speed	160 km/h (100 mph)
CO$_2$ emissions	< 40 g/km

The Flextreme concept from General Motors Europe has a double set of responsibilities: not only does it need to signal a refreshed European design direction for Opel, capitalizing on the momentum of the GTC Coupé (*The Car Design Yearbook 6*), but it must also act as a visual ambassador for GM's advanced E-Flex suite of low-emission powertrain technologies.

The E-Flex series began with the radically styled Chevrolet Volt at the 2007 Detroit show, setting the bar high for the European concept. And the Flextreme does not disappoint: its shape is elegant, well proportioned, and with interesting sculpted details. The layout is novel, too, with rear-hinged rear doors and a very innovative tailgate arrangement that sees the complete left and right rear quarters of the body hinge upward, pivoting around the center line of the roof.

Despite this feat of engineering, which gives remarkable access to the load area, black pillars and a black-glass roof allow the Flextreme's upper architecture to remain pleasantly light. A ridge runs down the sharply raked A-pillars and into the hood, to form the inner edges of the distinctive boomerang-shaped headlights; these very graphic units curl round the whole profile of the front from above the wheel arch to the level of the lower air intake.

The Opel grille is kept small and smooth, as if to symbolize the low cooling demands of the compact diesel engine, which, running at its peak efficiency point, drives the generator that powers the battery and the electric motors. At no point are the wheels directly driven by the engine.

The interior is strictly four-seater, with the rear floating seats set closer to the center to improve forward view; the driver has a single-spoke steering wheel and instruments directly ahead as well as across the full width of the windshield base. Something of a show gimmick is the pair of Segway scooters stowed in a compartment behind the rear apron.

Opel/Vauxhall Meriva

Design	Mark Adams
Engine	1.4-liter 4-cylinder turbo
Installation	Front-engined/front-wheel drive
Brakes front/rear	Discs/discs
Front tires	235/35R19
Rear tires	235/35R19
Length	4220 mm (166.1 in.)
Width	1760 mm (69.3 in.)
Height	1601 mm (63 in.)
Wheelbase	2640 mm (103.9 in.)
Track front/rear	1560/1584 mm (61.4/62.4 in.)

Though billed by Opel parent General Motors as a concept, this Meriva study is said by GM insiders to be close to the eventual production model. Significantly, too, it is expected that the production model will include the innovative Flex doors of the concept: the front doors hinge conventionally, while the back ones are rear-hinged.

This is important news in the small MPV segment that the Meriva will contest. The design moves the game along in terms of family-friendly travel: children getting out of the back seat descend from the car in the protected space between the front and rear doors, and with both doors opening through 90 degrees, that space is a good one.

Previously, rear-hinged doors had been abandoned on safety grounds for fear that they would swing open while driving: this risk, says GM Europe head of design Mark Adams, is completely eliminated in the Meriva thanks to an electronic interlock that allows the door to be released only when the car is stationary. In addition, says Adams, the B-pillar has been retained to ensure good side-impact protection.

However, there is more to the Meriva's design than just a clever door system. It is larger and much more stylish than today's plain model, and manages to exude a sense of sportiness despite its relative height. This is achieved thanks to a beltline that rises from the base of the A-pillar, takes a step down just aft of the B-pillar, and sweeps up into the triangular C-pillar. A second feature line runs forward from the apex of the rear lights, through the door handles, and then swings downward behind the front wheel arch to form a shallow recess in the door panel.

The rear is fun-looking, too, with the tailgate glass descending between the rear lights to form the license-plate surround.

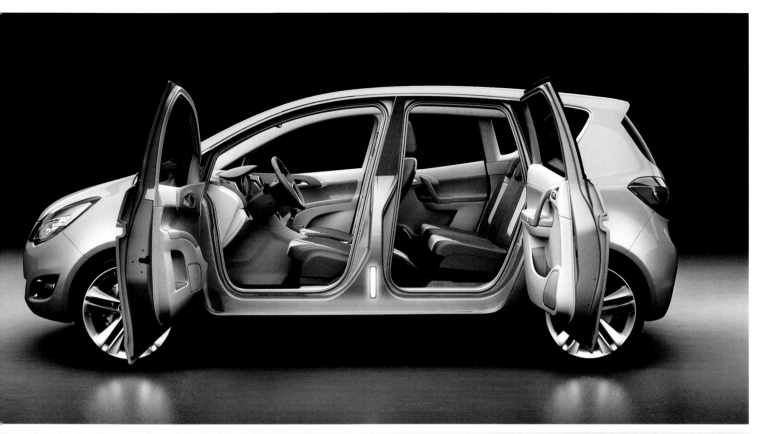

Peugeot 308

Design	Keith Ryder
Engine	1.6 in-line 4 (1.4 gasoline, and 1.6 and 2.0 diesel, also offered)
Power	112 kW (150 bhp) @ 5800 rpm
Torque	244 Nm (180 lb ft) @ 1400–4500 rpm
Gearbox	5-speed manual
Installation	Front-engined/front-wheel drive
Front suspension	MacPherson strut
Rear suspension	Torsion beam
Brakes front/rear	Discs/discs
Length	4276 mm (168.3 in.)
Width	1815 mm (71.5 in.)
Height	1498 mm (59 in.)
Wheelbase	2610 mm (102.8 in.)
Track front/rear	1536/1531 mm (60.5/60.3 in.)
Curb weight	1322 kg (2915 lb)
0–100 km/h (62 mph)	8.8 sec
Top speed	214 km/h (133 mph)
Fuel consumption	7.1 l/100 km (33.1 mpg)
CO$_2$ emissions	167 g/km

For the first of the 08-series Peugeots—models that are tasked with hauling the company back to proper profit levels in the next decade—the new 308 hatchback shows a surprisingly cautious approach to the evolution of the maker's house style.

Essentially, the 308 builds on the success of the outgoing 307, moderating its high build into a semi-tall architecture, and grafting on the design details and general ambience of the smaller, highly successful 207. The result is a car that is in many senses an enlarged 207; indeed, but for the fact that the 207 is a more harmonious overall design, the 308 could easily be mistaken for its smaller sister.

All the characteristic Peugeot elements are present: the full-width gaping lower grille mouth; the long, pointed headlamps; and the make's lion logo in exaggerated proportions, carried at the front apex of the V-shaped raised hood section that flows into the A-pillars. Because the screen is so fast—steeply raked backward—the point where the screen meets the roofline is a long way back, forcing the use of a large quarter-light on the front door.

The side is distinguished by a feature line that flows into the rear lights—though not at their points— and a rising beltline that helps to add visual mass to the rear. The rear screen has a mild wraparound and is sloped forward to give a sense of movement, while the horizontal feature line on the tailgate visually widens the car.

Inside, the focus has been on achieving quality of finish rather than originality of design. The smooth, soft-feel surface of the angled dash is punctured by five circular air vents, and a neat hooded instrument cluster is set in front of the driver. Even the center stack, paler in color, is kept clean, elegant, and simple.

Peugeot 308 RC Z

Engine	1.6 in-line 4
Power	163 kW (218 bhp)
Torque	300 Nm (221 lb ft)
Gearbox	6-speed manual
Installation	Front-engined/front-wheel drive
Front suspension	MacPherson strut
Rear suspension	Torsion beam
Brakes front/rear	Discs/discs
Front tires	245/40R19
Rear tires	245/40R19
Length	4276 mm (168.3 in.)
Width	1840 mm (72.4 in.)
Height	1320 mm (52 in.)
Curb weight	1200 kg (2645 lb)
0–100 km/h (62 mph)	7.0 sec
Top speed	235 km/h (146 mph)
CO$_2$ emissions	160 g/km

Extravagant and voluptuous, this 2+2-seater coupé concept may carry the 308 name, but it is a far cry from the humdrum 308 hatchback: the bodywork is entirely different, the tracks are wider, and the engine and suspension have been substantially uprated.

First impressions of the RC Z are of exaggerated and strange proportions. It is a car that initially appears overbodied, as if artificially elongated: the cabin is very small indeed, and the extreme slope to the rear of the glasshouse makes the rear end seem even longer, shifting the car's visual mass forward. In this, the design recalls the Volkswagen Karmann-Ghia coupés of the 1950s and 1960s.

Some have commented that the sweeping arch formed by the polished-aluminum cant rail looks inspired by the first-generation Audi TT; in contrast to that design, however, the RC Z is not as well resolved or harmonious. In particular, there is an awkwardness at the rear and in the angle between the flat rear deck and the double-bubble rear screen, which, because of its complex shape, has to be made of polycarbonate rather than glass. An intriguing detail is the rear muffler box, which doubles as an aerodynamic diffuser.

The vast gape of the grille occupies almost the whole of the front; at the side, the pronounced shoulder over the rear wheels, reinforced by a sudden step-up of the window line around the B-pillar, accentuates the forward-leaning, about-to-pounce stance of the car. Inside, perhaps disappointingly, the layout and fittings are taken straight from the 308 hatchback, though with the addition of leather trimming for the fascia panel and a Bell & Ross clock in place of the central air vent.

All in all, the 308 RC Z is a much more realistic proposal than many Peugeot sports concepts; cleverly exaggerated in its proportions, it is a flamboyant statement and, if Peugeot dares to build it, could succeed commercially.

Pininfarina Sintesi

Design	Giuseppe Randazzo
Engine	Hydrogen fuel cell
Power	160 kW (215 bhp) @ 2000 rpm
Gearbox	None
Installation	Electric motors in each wheel
Front tires	245/40R20
Rear tires	285/35R21
Length	4794 mm (188.7 in.)
Width	1988 mm (78.3 in.)
Height	1298 mm (51.1 in.)
Wheelbase	2990 mm (117.7 in.)
Track front/rear	1726/1676 mm (68/66 in.)
0–100 km/h (62 mph)	9.1 sec
Top speed	222 km/h (138 mph)

The Sintesi was greeted as perhaps the best of the concept vehicles at the 2008 Geneva show, as befits a design house as exemplary in its art as Pininfarina.

Long and low, it has a graceful, flowing form that sweeps smoothly from its delicate nose to its vertically chopped-off tail: nothing intrudes on the overall shape—no spoilers, air scoops, or excess detailing. A gentle wave motif links the front, side, and rear: at the nose, the base of the air intake subtly rises in the center, a device repeated at the rear to lighten the visual mass; at the side, with the doors open, the sill makes a pronounced double wave. The front doors open upward and forward; the rear ones pivot backward. Each contains a small opening window set into the main glass. At the back, the vertical rear glass is inset into the rim formed by the roof and the C-pillars, the recess sweeping round underneath to form the bumper area.

Yet the Sintesi is much more than just a beautiful shape. It uses hydrogen fuel-cell power to excellent effect in what Pininfarina calls "liquid packaging." The fuel-cell stacks and batteries are packaged along the long central chassis spine, leaving plenty of cabin length for the four occupants. The wheels are driven by individual motors, removing the need for a bulky transmission system.

Further innovation is to be seen in the interior, where a conventional dashboard is replaced by a glowing white module that sprouts from the center of the bulkhead and carries the instrumentation, the navigation and climate controls, and the steering wheel. The honeycomb support structure can be glimpsed through its translucent surface, and the steering wheel again echoes the wave theme in its lower edge.

Overall, the Sintesi is one of the most impressive concept cars ever to have featured in *The Car Design Yearbook*, and we expect that it will come to be seen as something of a milestone in design.

Pontiac Vibe

Design	Ron Aselton
Engine	2.4 in-line 4 (1.8 also offered)
Power	118 kW (158 bhp) @ 6000 rpm
Torque	224 Nm (165 lb ft) @ 4000 rpm
Gearbox	5-speed manual
Installation	Front-engined/all-wheel drive
Front suspension	MacPherson strut
Rear suspension	Double wishbone
Brakes front/rear	Discs/discs
Front tires	215/45R18
Rear tires	215/45R18
Length	4365 mm (171.9 in.)
Width	1765 mm (69.5 in.)
Height	1559 mm (61.4 in.)
Wheelbase	2600 mm (102.4 in.)
Track front/rear	1515/1495 mm (59.6/58.9 in.)
Curb weight	1495 kg (3296 lb)
Fuel consumption	10.2 l/100 km (23.1 mpg)

With the shift to the new-generation 2009 model year Pontiac Vibe, the car moves to a subtly different marketing segmentation: it's now being styled by parent company General Motors as a small crossover, whereas the outgoing model was seen as a sporty though practical hatchback.

The difference is immediately evident in the Vibe's bold new look: it is tough, chunky, and vibrant, and has a solid, muscular personality that is rare in a car that, for the United States at least, is regarded as small.

As before, the Vibe shares its architecture with the Toyota Matrix (itself also going into a new generation) and is built at the joint GM–Toyota plant in California—a sure guarantee of quality and reliability. The Pontiac brand book provides design features that work well on such a small model as this: the classic Pontiac twin grilles sit authoritatively at the front, just above a horizontal lower air intake, itself also finished with a purposeful mesh pattern.

The strong, clean, and easy-to-understand body features are defined by such key styling lines as the central spine running up the hood, the lines that run up from the front bumper and through the head-lamps to drop below the A-pillar, and the more subtle feature line running forward at door-handle level from the rear lights.

Pontiac has succeeded in achieving a distinctive DLO despite the high hood and the need for a step-down around the A-pillar; the C-pillar profile is strong, too, leading into a tailgate that is disappointingly plain, though protected by a solid-looking bumper surface drawn down from the tacky-looking rear lights.

Seen against the quality of the exterior design, the interior—shared with Toyota—is a further disappointment. Though the Toyota connection is a guarantee of top materials and manufacture, the fussy dashboard architecture, with too many isolated elements and functions, is a letdown.

Renault Kangoo

Design	Patrick le Quément
Engine	1.6 in-line 4 (1.5 diesel also offered)
Power	78 kW (105 bhp)
Torque	148 Nm (109 lb ft) @ 3750 rpm
Gearbox	5-speed manual
Installation	Front-engined/front-wheel drive
Front suspension	MacPherson strut
Rear suspension	Torsion beam
Brakes front/rear	Discs/discs
Front tires	195/65R15
Rear tires	195/65R15
Length	4213 mm (165.9 in.)
Width	1829 mm (72 in.)
Height	1799 mm (70.8 in.)
Wheelbase	2697 mm (106.2 in.)
Track front/rear	1521/1533 mm (59.9/60.4 in.)
Curb weight	1398 kg (3082 lb)
0–100 km/h (62 mph)	13 sec
Top speed	170 km/h (105 mph)
Fuel consumption	7.9 l/100 km (29.8 mpg)
CO_2 emissions	191 g/km

Renault invented a new market segment with its once-utilitarian Kangoo: basically a van design, the model soon found parallel popularity as a rugged, practical, and quirkily attractive no-frills family wagon—the kind of territory once occupied by the same firm's 4L in the 1960s and 1970s.

As a reflection of the growing importance of the sector, the second-generation Kangoo sees the idea grow up into something bigger and more sophisticated. The platform is now that of the larger Scenic MPV, the body is a full 7 inches longer, and the expression—at the front, at least—is that of the chic Twingo supermini. Even such refinements as cruise control and automatic lights and wipers are included—a far cry from the vanlike original.

In design terms the Kangoo is thoroughly contemporary, with a cheerful look stemming from its friendly water-droplet-shaped headlamps, its docile upright posture, and its generally toylike lack of sharp edges and corners. Even the windows have softly rounded corners, lending a playful air despite the substantial size of the combined cabin and luggage space.

Straight, upright sides are necessary to allow the sliding rear doors to open effectively, but, notwithstanding this design handicap, Renault has succeeded in making the Kangoo look strong through clever use of slightly recessed side windows (adding depth) and blistered wheel arches, which, with their crisp fold lines, catch the light and make for a clear differentiation between the surfaces.

At the rear the offset number plate is a quirky touch, and the very tall, pointed taillights on the D-pillars emphasize the height of the load compartment inside. The interior itself is playful, carlike, and fun, while further clever features include aircraft-style stowage compartments for rear passengers and built-in longitudinal roof rails that can swivel through 90 degrees to form a luggage rack capable of carrying loads of up to 176 pounds.

Renault Kangoo Compact Concept

Design	Patrick le Quément
Engine	1.5 in-line 4 diesel
Power	78 kW (105 bhp)
Torque	240 Nm (177 lb ft) @ 2200 rpm
Gearbox	6-speed manual
Installation	Front-engined/front-wheel drive
Front tires	255/50R19
Rear tires	255/50R19
Length	3896 mm (153.4 in.)
Width	1856 mm (73.1 in.)
Height	1717 mm (67.6 in.)
Curb weight	1225 kg (2700 lb)
0–100 km/h (62 mph)	11.5 sec
Fuel consumption	5.3 l/100 km (44.4 mpg)
CO2 emissions	139 g/km

Presented, to the surprise of many commentators, at the same time as the new-generation Kangoo five-door, this proposal for a shorter, sportier, and more youthful Kangoo is likely eventually to become the basis for the more compact half of a diversified and expanded Kangoo lineup.

The Compact Concept is built around the theme of young people's sports, specifically in-line skating. However, the only actual roller-skate-specific feature on the vehicle is the drop-down tailgate, which incorporates sockets to lock two pairs of in-line skates into position. The rest of the design is devoted to the twin ideals of looking good and enjoying outdoor life to the full.

The Compact Concept certainly looks fun, with its big, chunky wheels and tires, its bold frontal smile, and a range of colors that would do credit to a Fisher Price toybox. The tall, airy glasshouse is almost entirely glazed, its only visible structure being a single bar stretched between the cant rails above the B-pillar, itself slender and disappearing. With the rear roof section removed, it is as open as a pickup. The rigors of crash testing would surely call for a more comprehensive structure.

Color and texture contrasts are used to great effect: while the body sides, the pillars, and the cant rails are all finished in a vivid fireball orange, the hood and tailboard come in brushed aluminum; the lower air intake, in aluminum, has a solitary orange bar passing through it, as if this were a structural element holding the car's sides together.

The interior is an equally vivid mix of bright blue, smooth pale gray, and a selection of slip-proof silicone materials from the world of in-line skating; the rear bench seat can swivel through 180 degrees to face the back. This is Renault's equivalent of Citroën's C3 Pluriel—and Renault seems to have done a rather better job.

Renault Koleos

Design	Patrick le Quément
Engine	2.0 in-line 4 diesel (2.5 gasoline also offered)
Power	127 kW (170 bhp) @ 3750 rpm
Torque	360 Nm (265 lb ft) @ 2000 rpm
Gearbox	6-speed manual
Installation	Front-engined/four-wheel drive
Front suspension	MacPherson strut
Rear suspension	Multi-link
Brakes front/rear	Discs/discs
Length	4520 mm (178 in.)
Width	1850 mm (72.8 in.)
Height	1690 mm (66.5 in.)
Wheelbase	2690 mm (105.9 in.)
Track front/rear	1545/1545 mm (60.8/60.8 in.)
Curb weight	1645 kg (3627 lb)
Fuel consumption	7.9 l/100 km (29.8 mpg)
CO$_2$ emissions	209 g/km

The Koleos, the first SUV to be marketed by Renault, is another production model that has remained close to the concept that preceded it, though in this case the 2006 study, labeled a concept, was quite plainly very ready for manufacture.

The Koleos was developed by Nissan within the Renault network and is manufactured by Renault Samsung in Korea, but it quite clearly represents a Renault take on the modern idea of a crossover SUV. Though it incorporates all the classic SUV attributes of high ground clearance, a high seating position, and rugged detailing, including metallic skidplates, Renault has gone to some lengths to soften the image of a vehicle type often seen as aggressive and antisocial. The lines are soft and smooth rather than rigid and chunky, the window line plunges toward the front in a sporty fashion, and the strongly sloping tailgate is reminiscent of a big family hatchback. The height of the body is broken up by a feature line running rearward from the front wheel arch, decreasing the car's apparent visual mass.

The technique is repeated inside, where the Koleos opts for smooth, generally pale materials and clean surfaces to provide a family-friendly feel that is very different from the technical interior environments of many SUVs. The only visible concession to the 4x4 enthusiast is a centrally mounted display strip giving useful information on lateral and longitudinal slope angles and on the direction in which the front wheels are turned. The display can also call up altitude, compass, and barometric readings, and the Koleos's credentials are confirmed by its inclusion of a hill-start assist function and hill-descent control, the latter limiting downhill speed on steep slopes to about 4.5 mph.

Most customers, however, are likely to make greater use of the model's practical features, which include a split tailgate, providing a useful seating bench for picnics and sporting events. As with the Nissan Qashqai, which shares the same underpinnings, the Koleos is available with either front-wheel or four-wheel drive.

Renault Laguna

Design	Patrick le Quément
Engine	2.0 in-line 4 gasoline (1.5 and 2.0 diesel also offered)
Power	125 kW (168 bhp) @ 5000 rpm
Torque	270 Nm (199 lb ft) @ 3250 rpm
Gearbox	6-speed automatic
Installation	Front-engined/front-wheel drive
Front suspension	MacPherson strut
Rear suspension	Torsion beam
Brakes front/rear	Discs/discs
Front tires	215/55R16
Rear tires	215/55R16
Length	4695 mm (184.8 in.)
Width	1811 mm (71.3 in.)
Height	1445 mm (56.9 in.)
Wheelbase	2756 mm (108.5 in.)
Track front/rear	1557/1512 mm (61.3/59.5 in.)
Curb weight	1467 kg (3234 lb)
0–100 km/h (62 mph)	9.2 sec
Top speed	220 km/h (137 mph)
Fuel consumption	8.9 l/100 km (26.4 mpg)
CO$_2$ emissions	210 g/km

With Renault having publicly declared its aim of becoming a top-three brand in terms of product quality, the new Laguna—the company's top mainstream model—was especially eagerly awaited. But while the new Laguna appears impressive in its thoroughness and its fit and finish, the consensus at the Frankfurt show was that the same could not be said for its external design.

Previous Lagunas have been discreetly stylish; this latest edition is commendable in some details but disappointingly bland overall, and arguments are still raging over the questionable aesthetics of perhaps the most important aspect of any car, its front end.

The front has a full-width lower grille that appears to jut out, chinlike; above this, the bumper moulding widens toward the outside, becoming very bulky and giving a curiously front-heavy look. The hood is plain apart from a central crease, while a discreet feature line begins at the front wheel arch, sweeps upward to take in both door handles, and begins to curve back down again to form the baseline for the taillights on the rear panel. All the while, the waistline is rising toward the rear. The only really crisp area of the sedan's design is the interplay between the rear window, rear three-quarter panel, and rear lights; the wagon's rear is more successful, with something of a BMW Touring to it.

Inside is where Renault has made the greatest strides. The dashboard is broad, smooth, and sweeping, with quality materials, classy instruments, and such attractive details as the pale wood inserts, the subtle color combinations, and the general air of calm sophistication.

Coming from a company that has made a virtue of progressive design, however, the cautious—not to say ponderous—Laguna must be judged a surprising disappointment in terms of exterior design. To succeed, it will need to make up for this with exemplary quality.

Renault Laguna Coupé Concept

Design	Patrick le Quément
Engine	3.0 V6 diesel
Power	172 kW (230 bhp)
Torque	450 Nm (332 lb ft) @ 1700–3800 rpm
Gearbox	6-speed automatic
Installation	Front-engined/front-wheel drive
Brakes front/rear	Discs/discs
Front tires	245/35ZR20
Rear tires	245/35ZR20
Length	4685 mm (184.5 in.)
Width	1964 mm (77.3 in.)
Height	1372 mm (54 in.)
Wheelbase	2710 mm (106.7 in.)
Track front/rear	1661/1693 mm (65.4/66.7 in.)
Curb weight	1685 kg (3615 lb)

Despite being handicapped by a color scheme that did it no favors, the Laguna Coupé Concept comprehensively outshone the brand-new Laguna sedan and wagon when all three made their international debuts at the 2007 Frankfurt show.

Finished in all-over white, with no brightwork and with the wide-spoked wheels picked out in near-black granite, the Coupé seemed at first glance to be the handiwork of a tuning company. But closer inspection revealed it to be something rather more special.

Renault's vision for a top-of-the-range grand tourer, the Laguna Coupé Concept is a classically elegant two-plus-two with a long, sweeping form and long front but short rear overhang. The body surfacing is accomplished, too, with a soft, sinewy line that runs from the edge of the grille, over the fenders, and slightly downward to meet the rear lights, where it rises again to form the spoiler in the center of the tail. A further, more subtle feature line runs along the side of the roof, down alongside the rear screen, and out to the trunk edge to form the raised spoiler section. The soft body surfaces are broken up by such highly distinctive features as the full-width grille and the complex LED headlight blocks, each drawn from the sedan but looking rather more effective on the coupé.

The side windows form a single pure arc around their top edge, while the doors hinge upward and forward, Lamborghini style. The interior is rather more futuristic than is suggested by the classical exterior, with an elegantly curving dashboard in a calm gray faux-leather material; there are even copper highlights on the leather and metal mesh on the floor.

Overall, though this appears to be a relatively simple design, in fact a great amount of care has been put into the surfacing to create just the right balance of sportiness and sophistication.

Renault Mégane Coupé

Design	Patrick le Quément
Engine	2.0 in-line 4
Power	147 kW (197 bhp) @ 5800 rpm
Torque	280 Nm (206 lb ft) @ 2600 rpm
Gearbox	6-speed manual
Installation	Front-engined/front-wheel drive
Brakes front/rear	Discs/discs
Front tires	245/35ZR21
Rear tires	245/35ZR21
Length	4514 mm (177.7 in.)
Width	1908 mm (75.1 in.)
Height	1371 mm (54 in.)
Wheelbase	2749 mm (108.2 in.)
Track front/rear	1658/1648 mm (65.3/64.9 in.)
Curb weight	1310 kg (2888 lb)
0–100 km/h (62 mph)	7.2 sec
Fuel consumption	6.5 l/100 km (36.2 mpg)
CO_2 emissions	154 g/km

By the common consent of designers and commentators at the 2008 Geneva Motor Show, the Mégane Coupé concept marks a return to form for Renault design. It is a compact coupé that will inform the look of a future production coupé, and, more importantly, it gives a few hints of what the new mainstream Mégane hatchback will look like.

The Coupé concept is undeniably exotic in its allure, with a long, smoothly rounded hood leading to a steeply raked windshield extending right up over a long, sweeping roof; the long DLO is drawn out gracefully, ending in a neat C-pillar that curves inward in plan to form the almost boatlike tail.

The distinctive front recesses the Renault lozenge into the hood, above a low-set rectangular mesh grille; big triangular headlamps have deep air intakes set below them. At the rear the theme is echoed ambitiously with taillights that are set into deep recesses cut in the overall smooth shape. Inside, bold use of bright reds contrasts with gray shades for a classy yet futuristic ambience; of particular note is the organic podlike instrument housing, influenced by product design rather than automotive conventions.

Most remarkable of all, however, is the Mégane's double gullwing door layout, one of the most complex seen on any concept or production car. First, the full-length side windows and their associated glazed roof panels hinge upward from the center line of the roof, then the lower door panels—again full-length—swivel upward and outward on large lever arms pivoting behind the back seat. It is an intriguing and attractive arrangement, leaving the whole side of the car open yet sheltered from the rain. It would clearly be hopelessly expensive for production and would pose problems when opening the doors in an underground parking lot where height is restricted, but that should not detract from the Coupé's encouraging message about the future of Renault design.

Roewe W2

This good-looking Focus-sized concept car from Roewe was given a positive reception at the 2007 Shanghai auto show, but complex dealings between Chinese motor manufacturers in the months since mean the model could be much more significant than was at first expected. In fact, the W2 could be the first Chinese car to be built in Europe.

Some explanation is due here. When MG Rover collapsed in 2005, Nanjing bought the UK factory, the MG name, the assets, and the equipment to build the models; earlier, the much larger Shanghai Automotive International Corporation (SAIC) had acquired rights to the larger Rover 75 model, which it began building in China under the Roewe 750 label. SAIC also began developing smaller models at its UK engineering base, and the W2 is the result of that work. The big shift came at the end of 2007, when SAIC took over Nanjing, and thus the UK factory, bringing with it the possibility of slotting in the much more modern W2 instead of the dated Nanjing models.

The W2 could play well in the small sports sedan segment. It has a squat, solid stance, born of a sleek DLO profile, tight-fitting wheel arches, and lean, flattish sides topped by an attractive feature line that sharpens as it moves toward the taillights. The pillars are unusually thick, perhaps indicating a determination to perform well in safety tests, but a full-length glass roof helps to lighten the interior.

The rear, with its Lexus-like relationship between the rear-screen glass and the high deckline, is better resolved than the front, where too many elements compete for attention. The wing motif of the grille is unsophisticated, being too clearly contrived from plastic overlays on the mesh background, and the three lower air intakes add too much distraction. Nevertheless, the headlamp shape is attractive and could blend well with a bolder corporate-style grille, such as that on the Roewe 750.

Saab 9-4X Biopower

Engine	2.0 in-line 4 flex-fuel
Power	224 kW (300 bhp) @ 5400 rpm
Torque	400 Nm (295 lb ft) @ 2600–5100 rpm
Gearbox	6-speed automatic
Installation	Front-engined/all-wheel drive
Front suspension	MacPherson strut
Rear suspension	Multi-link
Brakes front/rear	Discs/discs
Front tires	245/55R21
Rear tires	245/55R21
0–100 km/h (62 mph)	8 sec
Top speed	235 km/h (145 mph)
Fuel consumption	10.5 l/100 km (22.4 mpg)
CO_2 emissions	251 g/km (gasoline)

Saab has not had the best of experiences with SUVs or crossovers. The 9-3X concept shown in Detroit in 2002 came to nothing, and the brand was pushed by its owner GM into teaming up with Subaru and GM Trucks to put its emblems on two very un-Saab-like models to increase showroom traffic in North America. The idea of a big, hefty, and crude 4x4 is a million miles from Saab's long-cherished brand values.

The new 9-4X, which shares its architecture with the Cadillac Provoq, is at least more in tune with Saab's philosophy that vehicles should not be oversized or extravagant. In terms of style, however, it pays no more than lip service to the Swedish brand's traditional visual aesthetic, with the familiar three-part grille frontal treatment and a bid to replicate the wraparound windshield look by blacking out all but the rearmost pillars. There is also a hint of the helmet look of the 9-3X concept in the shape of the C-pillar, echoed in the side feature line that kicks up into the pillar.

Yet, though it may not be thoroughbred Saab through and through, the 9-4X is at least a calm, fresh, and modern design as well as one that avoids the excessively macho aggressiveness of many heavy-weight SUVs. The overhangs are short, the wheels are big and their arches accentuated, but most of the 9-4X's other design cues spell sports station wagon rather than SUV.

The classy look is enhanced at the rear by the narrow strip of lights connecting the main lamps, and by the exhaust outlets neatly blended into the lower apron.

Inside, the 9-4X is traditional Saab but with paler colors and a translucent icelike plastic applied to selected sections of the dashboard. In the cargo compartment an ingenious storage system is able to stack three pairs of skis and their poles on a fold-up cradle within the vehicle.

Saab 9-X BioHybrid

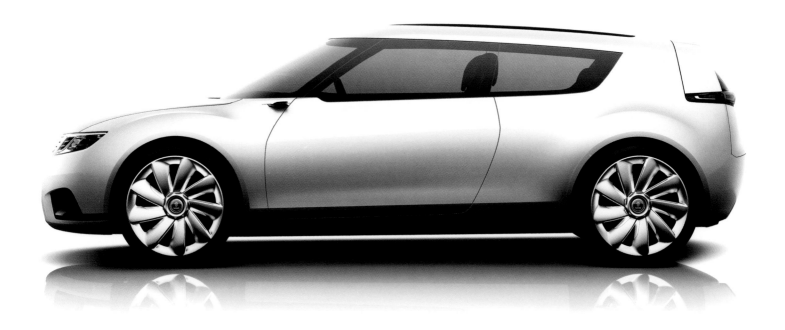

Design	Anthony Lo
Engine	1.4 in-line 4 and electric motor with lithium-ion battery pack
Power	149 kW (200 bhp) @ 5000 rpm
Torque	280 Nm (206 lb ft) @ 1750–5000 rpm
Gearbox	6-speed manual sequential
Installation	Front-engined/front-wheel drive
Front suspension	MacPherson strut
Rear suspension	Torsion beam
Brakes front/rear	Discs/discs
Front tires	245/35R21
Rear tires	245/35R21
Length	2705 mm (106.5 in.)
Width	1826 mm (71.9 in.)
Height	1382 mm (54.4 in.)
Wheelbase	4424 mm (174.2 in.)
Track front/rear	1610/1610 mm (63.4/63.4 in.)
0–100 km/h (62 mph)	7.9 sec
Top speed	215 km/h (134 mph)
Fuel consumption	6.4 l/100 km (36.8 mpg)
CO_2 emissions	105 g/km (on E85)

Saab has been on the point of launching into a small-car program for several years: the 9-X BioHybrid, shown at Geneva in 2008, is a strong pointer to the likely style of the eventual production car. That car will be a competitor in the market place against the likes of the Audi A3 and the BMW 1 Series, but on the road and in the showroom it will appear to be a radically different type of vehicle.

The 9-X BioHybrid is much more coupé than hatchback in style, but what makes it truly distinctive is the profile of the superstructure: the DLO tapers toward the rear, like an extended helmet visor, and the strongly wraparound windshield has Saab's characteristic beetle brow, which gives an enclosing, protective feel to the cabin. While the front strengthens the familiar Saab design elements of a three-piece grille and smoothly integrated headlights, it also introduces a new element, in the shape of deep lateral air scoops.

Many elements play on Saab's aircraft heritage, including the smoothly surfaced fuselage sides uninterrupted by door handles or other details, and the wheel design reminiscent of a jet engine's compressor fan.

The rear treatment marks new territory for Saab: the roofline continues horizontal right to the rear of the car and the very broad C-pillar and vertical tailgate glass. Below the glass is a slim rear-light strip, and below that an elegantly surfaced rear panel that successfully minimizes what could otherwise have been a heavy visual mass.

The interior reflects further progress in Saab's driver-centered thinking. Unusually, the driver areas are finished in white, with the passenger's side in dark gray. The front seats are white, the rear gray, again keeping the focus up-front.

Thoughtful details abound, including an active aerodynamic package with a moving underbody diffuser and a rear spoiler that extends at speed and tilts upward under heavy braking to give downforce over the back wheels and increase the visibility of the high-level brake light.

Scion Hako

Design	Thomas Bergeron
Installation	Front-engined/front-wheel drive
Brakes front/rear	Discs/discs
Front tires	225/45R18
Rear tires	225/45R18
Length	3701 mm (145.7 in.)
Width	1730 mm (68.1 in.)
Height	1461 mm (57.5 in.)
Wheelbase	2400 mm (94.5 in.)

A broad mix of designs has emanated from Toyota's American youth brand, Scion, in the few years that it has been in the US market. Concept cars and production models have ranged from the very ordinary to the frankly absurd, and this latest, the Hako, slips in toward the latter end of that spectrum.

Shimmering in its bright orange paintwork and burnished oversized alloy wheels, it looks at first like a low-riding product of America's thriving West Coast hot-rod and customizing culture. There is a bluff nose, rounded in side profile but with a small inset rectangular black grille; this rises to an old-style flat hood, with the equally retro extended, flat-topped wheel arches. Narrow slit-eye LED headlamp blocks squint forward perplexedly.

But strangest of all, and dominating the design from almost any viewing angle, is the windshield. It is completely flat and completely vertical, just like that of a chopped hot rod; yet it also wraps round upward through 90 degrees to blend into the roof, which just happens to be yet another flat surface.

From the upright A-pillars the cant rail runs straight back, then travels round the wraparound rear window (which is also vertical, but curved in plan) to return to the pillar at the other side. All other pillars are blacked out, giving the impression of a roof cantilevered off nothing but the front pillars.

Two long doors give access to the interior, which is a typically West Coast techno-fest of interactive camera systems, joystick-style controls, and video screens. Fish-eye camera lenses mounted externally project deliberately distorted images onto the Hako's numerous display screens and, says Scion, interface with the music systems.

How one responds to the Hako must, in the end, be a matter of personal taste. Its form, its function, and even its future are unclear. To some, it might be the height of cool, but to the rest of the world it is just too crazy for comfort.

SEAT Bocanegra

Design	Luc Donckerwolke
Engine	1.4 in-line 4
Gearbox	7-speed dual clutch
Installation	Front-engined/front-wheel drive
Brakes front/rear	Discs/discs
Front tires	235/35R19
Rear tires	235/35R19
Length	4063 mm (160 in.)
Width	1712 mm (67.4 in.)
Height	1350 mm (53.1 in.)
Wheelbase	2461 mm (96.9 in.)
Track front/rear	1477/1477 mm (58.1 in.)

Displayed as a concept at the 2008 Geneva show, the Bocanegra—"black mouth" in Spanish—serves as a taster for the new-generation Ibiza hatchback. The two show a striking similarity in overall look, with only the more extravagant detailing of the Bocanegra marking it as a show car.

True to its name, the Bocanegra impresses with its large black face, with a black central grille and a broad black undergrille flanked by black-meshed lateral brake-cooling air scoops. Even the headlamps hide behind black Perspex, and the area between the lights and grilles is black-finished, too. The contrast with the fiery red bodywork makes it look as if the car has been dipped nose-first into dark chocolate.

The sporty front sets off a style that is equally youthful and racy. The hood line formed over the headlights sweeps backward and plunges down into the door; another sharp crease creates a shelf over the rear wheel arch, adding a dynamic element to the rear. The interplay of these lines and surfaces creates a lively, active look, reinforced by the tight wheel arches and the wide stance of the five-spoke wheels. From the rear the Bocanegra spells hot-hatch, with a spoiler integrated above the rear window and an echo of the frontal style in the shape of a central diffuser outlet under the bumper flanked by meshed grilles at either side.

A further extension of the black/red theme is to be found inside, where the unusual quilted black seats have red diagonal stitching and red side bolsters; there are also red accents on the carpets and door trims, and a smart, glossy piano-black inset panel running most of the width of the dashboard. This mixing of gloss and matt textures adds interest to the interior.

While the more intense detailing of the Bocanegra will not translate into production, the concept is a welcome sign that SEAT's design language is progressing out of the rut worn by the unattractive Altea and Toledo.

SEAT Tribu

Design	Luc Donckerwolke
Installation	Front-engined/four-wheel drive
Brakes front/rear	Discs/discs
Front tires	255/50R20
Rear tires	255/50R20

It is no secret that SEAT, the Spanish outpost of the Volkswagen group, has been struggling in recent years. Its oddly styled hatchbacks and sedans (reviewed, often unfavorably, in successive editions of *The Car Design Yearbook*) have conspicuously failed to find as much favor with customers as the models of VW's other value brand, Skoda.

The Tribu—at present just a concept—represents two facets of the brand's fightback: entry into new market segments (in this case, that of a compact SUV) and the presentation of a refreshed visual identity developed under new design director Luc Donckerwolke.

A striking, compact off-roader, the Tribu is sporty, bold, and fun-looking; in this sense, it annexes similar ground to that of the original three-door Toyota RAV4 in the 1990s. Powerful design cues affirm its role as a tough-terrain crosser: the bulging, squared-off wheel arches have huge voids to display the generous wheel articulation, while strong geometric forms combine to create a distinctive body shape.

The front end is highly expressive, a bold trapezoidal central grille being flanked on either side by equally large five-sided grilles, each topped by inset double-headlamp units. Sharp body creases bisect large taut surfaces, the hood being one example. As a frontal identity for future mainstream SEAT models, it would work well.

The all-glass hatchback has a novel mechanism that allows the tailgate to open in two phases, opening partially at first, and then rolling up into the roof. Inside, there are once again strong geometric shapes and the now-familiar combined windshield and glass roof.

Built on the same 4x4 platform as VW's Tiguan and Audi's upcoming Q3, the Tribu also incorporates an electronic drive mode selector that adapts the engine, suspension, and transmission to the prevailing conditions. Fun, different, and distinct, the Tribu is just the kind of car that could rekindle interest in SEAT.

Skoda Superb

Design	Jens Manske
Engine	3.6 V6 (1.4 and 1.8 in-line 4, and 1.9 and 2.0 diesel, also offered)
Power	191 kW (256 bhp) @ 6000 rpm
Torque	350 Nm (258 lb ft) @ 2500–5000 rpm
Gearbox	6-speed automatic
Installation	Front-engined/all-wheel drive
Front suspension	MacPherson strut
Rear suspension	Multi-link
Brakes front/rear	Discs/discs
Front tires	225/45R17
Rear tires	225/45R17
Length	4838 mm (190.5 in.)
Width	1817 mm (71.5 in.)
Height	1462 mm (57.6 in.)
Wheelbase	2761 mm (108.7 in.)
Track front/rear	1537/1510 mm (60.5/59.5 in.)
Curb weight	1665 kg (3671 lb)
0–100 km/h (62 mph)	6.5 sec
Top speed	250 km/h (155 mph)
Fuel consumption	10 l/100 km (23.5 mpg)
CO₂ emissions	238 g/km

It is a reflection of Skoda's growing self-confidence and its secure position in the market that the second generation of its Superb upper-medium model is a much bolder and more individual design. In particular, it unashamedly projects forward its big new-look Skoda grille, reflecting the greater pride in the brand. The outgoing Superb, by contrast, was little more than a warmed-over previous-generation Passat.

But while the new model still takes its underpinnings from the Passat—albeit the newer model with the transverse engine—its design carries little that could visually link it to the Volkswagen. The front end contrives a strong and classy premium look thanks to the smart grille thrust forward from the intricately shaped headlamps; the hood carries a lot of contours, too, trailing rearward from the central Skoda emblem and from the grille corners at each side. A big, full-width lower grille complements the vertical louvres of the upper signature grille.

There is less obvious interest to the sides, though the long cabin and the door openings cut close to the rear wheel arch signal a spacious interior. The external style begins to lose its identity toward the rear, with thick, kinked C-pillars reminiscent of BMW treatment, an abbreviated trunk lid, and large rear lights that fail to match up to the classy message of the frontal style.

Inside, however, the Superb is likely to delight its customers. The extended wheelbase gives it space aplenty, and the smooth, harmonious dashboard is clearly influenced by Audi for a clean, stylish, and sophisticated look. The soft-touch plastics on the dash and center console are of clearly good quality: only the bulky, messy steering wheel design comes as a disappointment.

The main innovation on the Superb is the so-called Twindoor trunk, an arrangement that ingeniously allows the trunk lid to be opened conventionally like a normal sedan's, or to be raised as part of the tailgate, as on a hatchback.

Subaru Forester

Engine	2.4 flat 4
Power	167 kW (224 bhp) @ 5200 rpm
Torque	307 Nm (226 lb ft) @ 2800 rpm
Gearbox	4-speed automatic
Installation	Front-engined/all-wheel drive
Front suspension	MacPherson strut
Rear suspension	Double wishbone
Brakes front/rear	Discs/discs
Front tires	225/55R17
Rear tires	225/55R17
Length	4560 mm (179.5 in.)
Width	1780 mm (70.1 in.)
Height	1674 mm (65.9 in.)
Wheelbase	2615 mm (103 in.)
Curb weight	1560 kg (3440 lb)

The hostile reception given to the bland look of Subaru's new-generation Impreza last year has not deterred the Japanese four-wheel-drive specialist from persisting with its current design policy, which appears to have caution and inoffensiveness at its heart.

It is thus that the new Forester, essentially a more off-road-oriented spin-off of the Legacy station wagon, has become taller and more SUV-like in its bid to differentiate itself more clearly from the Legacy. In so doing, however, the Forester has become more anonymous, a me-too focus-group pleaser.

It is a design that works better in darker colors than in the plain silver of the Detroit show car; the silver is cruel in revealing the lack of interest on the vehicle's sides—apart from the curious double eyebrows above each wheel—and also has the effect of hiding the angle on the front of the hood where it dips to meet the grille. Again, the chrome grille looks stronger and wider on the darker car.

The side view clearly shows how Subaru has neatly tucked in the front and rear overhangs so as to allow good off-road ability; the extra height is most apparent from the rear, where the number-plate positioning and the shape of the rear lights tend to exaggerate the tall look.

The interior, again disappointingly, appears to be little more than a rerun of what appeared in last year's Impreza. The small steering wheel and the semicircular instrument pack in front of the driver are pleasant enough, but the gray-on-gray color scheme sets an uninspiring tone, and even the addition of double stitching and improved materials cannot do much to enhance the low-rent ambience.

Subaru clearly needs to improve the standard of its interior design; more importantly, it must realize that by setting out not to offend anyone, it risks not pleasing anyone either.

Subaru G4e

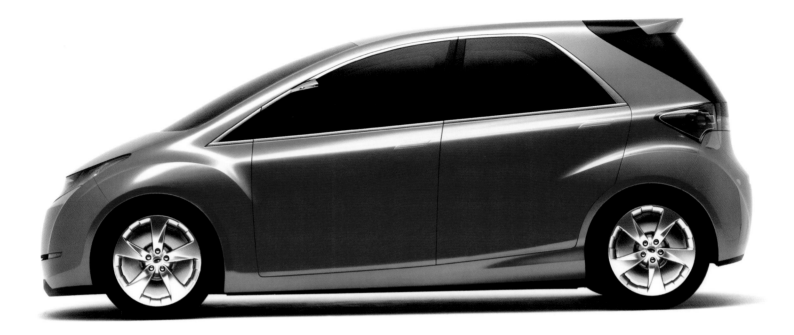

The Subaru G4e was a concept that attracted much attention at the 2007 Tokyo show, but the interest came more from analysts and technical specialists than from aesthetic designers. For while there is nothing particularly remarkable about the G4e's exterior, the headline figures relating to its powertrain and performance made engineers sit up and take notice.

Electric vehicles have always struggled with the intertwined burdens of excess weight, poor performance, and inadequate range: the culprit in all this has been the big, bulky, and weighty battery. Subaru claims that its next-generation version of the lithium-ion battery extends range further. Vanadium, says Subaru, is able to store two to three times more ions than conventional materials at the positive terminal, resulting in double the energy density of a manganese lithium-ion battery of the same weight.

This increased energy storage, allied with a lightweight structure, an aerodynamic shape, and the underfloor positioning of the battery, gives the G4e an expected range of 125 miles on a single charge—an impressive figure for a vehicle that can seat five in comfort. Subaru has revealed little else about the engineering of the G4e, but says that a fifteen-minute charge can fill the batteries to 80 percent capacity.

In terms of exterior design, the G4e's pyramid-like shape breaks little new ground, in either its proportions or its details. The rounded front has no grille, but the dark band linking the lights gives a half-hearted smile; the shoulder form gradually grows as it flows back through the doors and eventually provides the profile of the rear lights. The steeply sloping C-pillar clashes somewhat with the upright rear screen and, especially, the tacked-on body-color spoiler on the top.

However, the worst excesses are reserved for the interior, where garish red sits alongside brash gold and cream for a tacky boudoir effect, despite the innovative cascading shape of the display panel linking the windshield and the floor.

Engine	Electric motor—lithium-ion battery
Power	65 kW (87 bhp)
Length	3985 mm (156.9 in.)
Width	1695 mm (66.7 in.)
Height	1570 mm (61.8 in.)
Wheelbase	2650 mm (104.3 in.)

Subaru Impreza

Engine	2.5 flat 4
Power	167 kW (224 bhp) @ 5200 rpm
Torque	307 Nm (226 lb ft) @ 2800 rpm
Gearbox	5-speed manual
Installation	Front-engined/all-wheel drive
Front suspension	MacPherson strut
Rear suspension	Double wishbone
Brakes front/rear	Discs/discs
Front tires	205/50R17
Rear tires	205/50R17
Length	4580 mm (180.3 in.)
Width	1740 mm (68.5 in.)
Height	1475 mm (58.1 in.)
Wheelbase	2620 mm (103.1 in.)
Curb weight	1425 kg (3142 lb)

After the astonishing success of the iconic rally-inspired cars in the 1990s, Subaru has had something of a rough ride with subsequent Imprezas, at least as far as the design critics are concerned. The second-generation car was universally lambasted as bug-eyed and ugly, and gained respectability only later in life with a tidy-up facelift that took away much of its character.

So expectations surrounding the launch of the all-new 2007 Impreza were high, especially among enthusiasts hoping that the rally-replica WRX version would rekindle the glamorous image of the make. Yet when the standard Impreza was unveiled at the 2007 New York show, the reaction was overwhelmingly negative: the new car is as plain and pedestrian as its predecessors were racy and aggressive.

Perhaps the only distinct features of the basic model are the curved roof profile and the long feature line that travels forward from the wraparound rear lights. The rest is sensible but unremarkable; even the gray-on-gray interior is nondescript.

With little in the design to provide continuity with previous Imprezas, enthusiasts held their breath in the hope that the higher-performance models, revealed several months later, would supply the necessary aesthetic adrenaline. Once again, however, they were disappointed: despite such familiar rally hallmarks as the hood air scoop, the broad underbumper intakes, and the flared wheel arches, even the WRX Sti has failed to excite its target public.

It is unusual to see a respected carmaker get it so wrong when replacing its highest-profile product. The outgoing Impreza in its contrasting Jekyll-and-Hyde guises managed to please both the sedate Saturday shopper and the tire-smoking, 300-horsepower rally wannabe; with the new model, Subaru appears to have deserted the high-image, high-speed fraternity, and it will take a lot of clever styling makeovers to win them back.

Suzuki Concept A-Star

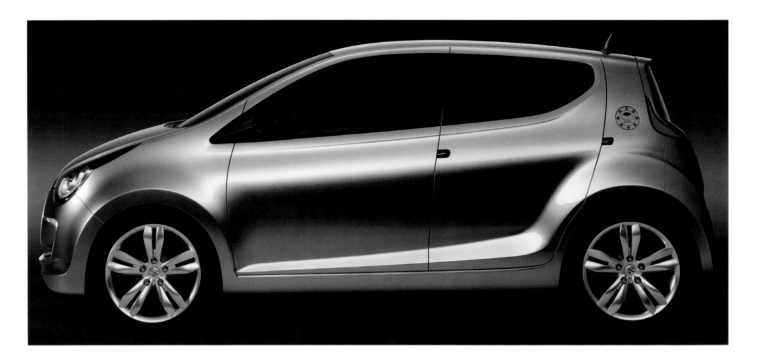

Design	Suzuki/Maruti Suzuki India
Engine	1.0 in-line 3
Gearbox	5-speed manual
Installation	Front-engined/front-wheel drive
Brakes front/rear	Discs/discs
Front tires	205/45R17
Rear tires	205/45R17
Length	3580 mm (140.9 in.)
Width	1680 mm (66.1 in.)
Height	1400 mm (55.1 in.)
Wheelbase	2360 mm (92.9 in.)
CO_2 emissions	109 g/km

Following the Swift, last year's Splash, and the more recent series of Kizashi large cars, this is a further Suzuki concept vehicle to have impressed showgoers in Europe. Each concept has shown definite style and originality, and now the A-Star promises to bring these same qualities into the small-car class, where Suzuki's production offerings to date have lacked distinction.

The A-Star has added significance in that it is jointly developed with India's Maruti Suzuki and will be built in big numbers in India for worldwide distribution, promising a strong price entry point in European markets. The A-Star is also tasked with giving Suzuki an ultralow CO_2 emissions vehicle for Europe, the target being a figure below the psychologically important 100 grams-per-kilometer barrier.

For a supereconomy car, the A-Star looks like a lot of fun, too. A compact five-door hatchback, it has short overhangs, prominent wheel arches, and a rising beltline, all of which give it a sporty stance. The semicircular headlamps have the look of eyes and have playful pink eyebrows, adding a human touch. Other pink accents—such as the door mirrors, the wheel hubs, and flashes in the front seats—impart a modern, fashionable feel, and the rear lighting, with elegant boomerang-shaped taillights and a neat central Suzuki logo that illuminates as the high-level brake light, is a real standout detail that could become as distinctive for Suzuki as the frontal light rings are for BMW.

Yet the A-Star is more than a set of details, however intriguing those details may be. The proportion and the surfacing of its body are expertly handled, especially the way in which the sill line rises to match the rising waistline to add interest to the vehicle's sides. The use of interior lighting is original, too: a diffused pink glow emanates from the gap between the gray upper layer of the dashboard and the pale lower section, with a similar glow illuminating the door pockets.

Suzuki PIXY & SSC

Branded as a "people-focused sustainable mobility vehicle concept," Suzuki's PIXY could be almost anything from a small, personal sports car to a carbon-neutral minibus with room for twenty.

However, the concept turns out to be something rather different, or, strictly speaking, two things that are rather different. PIXY is an individual personal mobility device, much along the lines of Toyota's i-Swing of 2005—a slightly reclined armchair cocoon running on four wheels and with a full-depth fold-forward Perspex bubble windshield to keep the weather out and the heat in. Portholes at eye level on the sides give some measure of visibility, but Suzuki says little about the device's performance on its hydrogen- and solar-energy supply and, in any case, the fact that the front wheels are casters will be a deterrent to going too fast.

The other half of the equation is SSC, standing for Suzuki Sharing Coach. Eschewing the PIXY's smooth curves for a straight-lined hexagonal theme, this narrow vehicle looks like a downsized ski lift but in fact acts as the docking station for a pair of the individual mobility devices. Both the narrow ends and the long sides of the SSC hinge open, allowing the PIXY to drive in; what is more, both ends are identical. Once the PIXYs are in, the SSC can be driven, tiny wheels at each corner propelling it, and the lights—again symmetrical—illuminating red or white, depending on the direction it is being driven.

The idea is for the SSC to transport the PIXYs to buildings, pedestrianized areas, or other traffic-free zones: the PIXYs then disembark, allowing their users to complete their journeys under hydrogen power rather than on foot. Suzuki has not yet suggested a solution that can cope with steps or escalators, but it did show sketches of two further PIXY carriers—a twin-hulled speedboat and a single-seater sports car. Tokyo 2009, here we come.

Suzuki Splash

Engine	1.2 in-line 4 (1.0 gasoline, and 1.3 diesel, also offered)
Power	63 kW (85 bhp) @ 5500 rpm
Torque	114 Nm (84 lb ft) @ 4400 rpm
Gearbox	5-speed manual
Installation	Front-engined/front-wheel drive
Front suspension	MacPherson strut
Rear suspension	Torsion beam
Brakes front/rear	Discs/drums
Front tires	185/60R15
Rear tires	185/60R15
Length	3715 mm (146.3 in.)
Width	1680 mm (66.1 in.)
Height	1590 mm (62.6 in.)
Wheelbase	2360 mm (92.9 in.)
Track front/rear	1470/1480 mm (57.9/58.3 in.)
Curb weight	990 kg (2183 lb)
0–100 km/h (62 mph)	12.3 sec
Top speed	174 km/h (108 mph)
Fuel consumption	5.6 l/100 km (42 mpg)
CO$_2$ emissions	136 g/km

For Suzuki Splash, read Opel/Vauxhall Agila. Or, strictly speaking, switch the order round, for this Suzuki is the vehicle from which Opel has commissioned its carbon-copy Agila design.

The original Splash concept, true to its name, did indeed make quite a splash when it was presented at the 2006 Paris show. Encouragingly, the eventual production car remains faithful to the concept's sense of style, fun, and originality. Though its emlem is Japanese and far from glamorous, this is a fashionable contemporary small car targeted at young people—a big contrast to its boxy and boring Wagon-R predecessor, which found favor almost exclusively with older folk.

In profile, the Splash is characterized by a roomy-looking arched-roof cabin, arrow-shaped rear lights wrapping round the sides of the C-pillars, and a near-vertical tailgate dropping down to suspiciously Smart-like rear wheels stuck so far back that their arches flow directly into the rear bumper.

At the rear the Suzuki mounts its license plate in the center of the tailgate, giving a more sedate look than on the Opel, where the plate is recessed into the bumper. The same holds true at the front, where the Suzuki comes across as more orthodox—and perhaps more Fiat-like—with its straight-bottomed grille and headlights; on the Opel-branded version these elements are curved at their bases, making for a more emotive expression in the "eyes" and "mouth." A strong bar of body color separating the lower and upper grilles prevents the front end looking too tall.

There is less that is interesting from the design perspective in the roomy and clean-contoured interior (which, incidentally, is identical to the Opel's but for the fabric choices), apart from the stylish and informative single central instrument, which houses a neat speedometer surrounded by a colorful arc of warning lights.

Suzuki X-Head

This tough, hunky, and dramatic-looking gray and lime-green device, with big, fat tires, car-crushing bumpers, and a no-nonsense upright driver's cab, is the perfect monster truck—until you realize it is only just over 12 feet long. So convincing is Suzuki's visual trickery that every detail comes across as authentic, generating the impression of a real, full-size Tonka Toy. Completing the illusion are the generous ground clearance, the voluminous square-cut wheel arches set for substantial suspension travel, the solid skidplate protecting the vulnerable front underbumper area, and the hefty strengthening bars triangulating the side between the rear of the cab roof and the top of the tailboard frame.

Yet the example shown by Suzuki is no idle plaything: a pair of Suzuki motocross bikes is lashed securely to its load bed; the drop-down sides house all the spare parts, helmets, and protective clothing needed for two-wheeled competition; and a powered tailgate helps with loading heavy machinery. Three different load beds are proposed: a camper, sleeping two; a rescue configuration (much like the Daihatsu Mud Master); and a variant described by Suzuki as fashion for stylish urban mobility.

Despite these gimmicky-sounding applications, there is much design interest to be found in the X-Head. LEDs are used for both headlights and taillights, the mix of gray panels inside and out adds a contemporary touch—as well as contrasting tastefully with the green body color—and an octagonal theme is picked up in several locations, notably the seats, the steering wheel, and the bank of air vents above the dashboard. Rugged-looking aluminum switchgear on the roof panel and the center console highlights the toughness of the concept and the skillful utilitarian design of the whole vehicle.

Who knows? A market might yet exist for a small, stylish, truck-like vehicle that is strong on mobility and durability but relatively light on costs and environmental impact.

Installation	Front-engined/four-wheel drive
Length	3800 mm (149.6 in.)

Tata Nano

Design	I.DE.A, Pierre Castinel, Tata
Engine	0.6-liter two-cylinder gasoline
Power	25 kW (33 bhp) @ 5000 rpm
Torque	48 Nm (35 lb ft) @ 2500 rpm
Gearbox	4-speed manual or CVT
Installation	Rear-engined/rear-wheel drive
Front suspension	MacPherson strut
Rear suspension	Coil spring
Brakes front/rear	Discs/drums
Front tires	R12
Rear tires	R12
Length	3100 mm (122.1 in.)
Width	1500 mm (59.1 in.)
Height	1600 mm (63 in.)
Track front/rear	1325/1315 mm (52.2/51.8 in.)
Top speed	75 km/h (47 mph)
Fuel consumption	5 l/100 km (47 mpg)

In terms of global public attention, the Tata Nano has been one of the highest-scoring car designs of recent times. Unsurprisingly, the main interest has centered on just how it has been possible to provide a safe and comfortable four-passenger vehicle selling for only US$2500. Many doubted that it would be practical at all, but Tata has proved the skeptics wrong with its classically simple curved monovolume shape, set on four small wheels (see feature page 24).

The constraints were many. The design had to accommodate four people and luggage, yet had to be kept short in order to minimize weight and material use, thus allowing lighter braking, steering, and suspension components, and all the parts had to slot together simply to streamline manufacturing.

The result is a compact 10-foot design, with the engine placed under the rear seat to maximize use of space, and with the arched passenger cabin taking up virtually the whole length of the vehicle. The profile is tall and narrow, and front and rear overhangs are minimal. Most of the side is made up of doors, but interest is added by a crisp feature line that runs back above the door handles, while an upward curve in the rear door helps to break up the height of the sides.

The near-vertical hood panel meets the flat windshield in one continuous arc, which carries through the arched roof to a clean, simple tailgate flanked by tall, upright rear lights. The rear bumper has slots to allow the exit of cooling air for the underfloor rear engine.

Simplicity again rules in the interior. An upright center stack combines the single round instrument dial with the twin air vents and the minor switchgear and heater controls; the main expanse of the dashboard is a full-width storage shelf.

The Nano does not aim to be elegant, sporty, or exciting. Instead, it is a clever and appealing solution that is likely in years to come to be seen as a landmark design.

Toyota 1/X

Engine	0.5-liter flex-fuel hybrid
Installation	Mid-engined/rear-wheel drive
Length	3900 mm (153.5 in.)
Width	1620 mm (63.8 in.)
Height	1410 mm (55.5 in.)
Wheelbase	2600 mm (102.4 in.)
Curb weight	420 kg (926 lb)

The Toyota 1/X—pronounced "1/Xth," according to the company—is the perfect illustration not only of the positive propaganda, but also of the perils that can come with choosing a high-profile international forum to exhibit what in effect is work in progress.

Widely touted as a template for the next-generation Prius—the current model is the world's biggest-selling hybrid and a flag-carrier for eco-conscious driving everywhere—the 1/X hit the headlines with some astonishingly green vital statistics, including an overall weight of 926 lb and CO_2 emissions expected to be half those of the already highly regarded Prius. These results are achieved by a complete rethink of the layout of the car and the materials from which it is made up: the engine, a tiny two-cylinder flex-fuel unit, is under the back seat, and futuristic lightweight carbon-fiber-reinforced plastic replaces metals for the car's structure and bodywork.

But anyone expecting the 1/X to be a breathtaking vision of the future on a stylistic level will be sorely disappointed. The design is not one of beauty: it is more of an engineer's car, its simple forms confining themselves to those that can be conveniently formed using the new materials. Nevertheless, it is clear that the superior strength of carbon fiber (it is used on racing cars) will allow designers to specify much narrower roof pillars, permitting improved visibility, and that the LED headlamps that fill the entire hood panel are a viable solution. The tires are very narrow, to reduce air drag and rolling resistance, which could also become a trend.

But without a proper interior, with no solution for the doors, and with no sound insulation whatsoever, the 1/X can in no sense be realistically compared with current models. Its role is as a structural technology demonstrator, not an aesthetic statement, and Toyota must be hoping the public will not make the mistake of judging it on appearances alone.

Toyota 1/X **Concept** 253

Toyota A-BAT

Design	Ian Cartabiano
Engine	In-line 4, gasoline/electric hybrid
Gearbox	CVT
Installation	Front-engined
Brakes front/rear	Discs/discs
Front tires	19 in.
Rear tires	19 in.
Length	4605 mm (181.3 in.)
Width	1890 mm (74.4 in.)
Height	1625 mm (64 in.)
Wheelbase	2850 mm (112.2 in.)

Sensing that environmental and fuel-cost concerns are making US consumers start to retreat from the macho world of the traditional heavyweight pickup truck, Toyota has entrusted the A-BAT concept with the mission of bringing into this redneck sector the softer, caring, ecological values that made its Prius hybrid such a hit.

For a pro-environmental model, the A-BAT looks surprisingly tough and muscular. Its high-waisted proportions, low roof, and heavy pillars impart an aura of solidity, reinforced by the angular theme to the body surfacing and the dark silver finish, which makes the design appear to be milled from a solid block of steel or aluminum.

When the car is viewed from the side, the feeling of height is accentuated by the substantial depth of body side above each wheel arch and by the half-height headlights visible from the side. The smooth, carlike shape of the cab, with its fast windshield sweeping into the low cant rail, contrasts with the blocky understructure and is set well forward, again creating an unusual proportion.

The front, again shaped out of heavy-looking metal sections, comes across as clumsy and poorly resolved, and certainly in no way suggestive of an eco-friendly powertrain underneath.

Opening the four doors to the interior provides another shock, with overwhelmingly yellow-clad furnishings marking out a polygon theme in the seat frames, the center console, the door panels, the dashboard, and the flat-topped steering wheel. While the folding rear seats are a practical aspect, allowing the load bed to be extended right into the cab, the overall feel of the interior design is concept-car gimmicky, and different just for the sake of it. Controls are simplified to the extreme, with just a steering wheel, two pedals, column stalks, and twin knobs on the center stack. The horizontal areas on top of the dashboard, at either side of the central spine, are taken up by big solar panels, able to charge the vehicle's batteries when it is parked.

Toyota Corolla

Engine	2.4 in-line 4 (1.8 also offered)
Power	118 kW (158 bhp) @ 6000 rpm
Torque	220 Nm (162 lb ft) @ 4000 rpm
Gearbox	5-speed manual
Installation	Front-engined/front-wheel drive
Front suspension	MacPherson strut
Rear suspension	Double wishbone
Brakes front/rear	Discs/discs
Front tires	215/45R17
Rear tires	215/45R17
Length	4529 mm (178.3 in.)
Width	1699 mm (66.9 in.)
Height	1486 mm (58.5 in.)
Wheelbase	2601 mm (102.4 in.)
Track front/rear	1514/1496 mm (59.6/58.9 in.)
Curb weight	1157 kg (2550 lb)
Fuel consumption	9 l/100 km (26.1 mpg)

Toyota is an acknowledged master of the art of conservative design, producing as a major part of its broad range the sort of nondescript vehicles that are liked by people who do not especially like cars.

Yet there is no denying the success of the strategy. By ensuring that its staple products do not offend anyone, Toyota may have earned the scorn of many in the design community, but it has also earned the loyalty of millions of customers worldwide; in fact, the Japanese giant is poised to overtake GM as the world number-one carmaker in 2008.

The Corolla, now entering its tenth generation and with more than thirty million examples sold since its debut in the 1960s, is the perfect illustration of this strategy in action. It is a middle-class car that has grown up with the age, the aspirations, and the affluence of its core clientele. In its latest global incarnation it is claimed to be lower and sportier than the previous model, but few would notice: it is neat, clean-lined, and roomy enough for four, which is all that anyone asks of it. Only if one strains the eye, looking along the car's flanks from the rear and disregarding the rather incongruous spoiler on the trunk lid, can the sporty forward thrust of the A-pillar be discerned; the C-pillar, likewise, is drawn further rearward in the interests of a sleeker style.

It is a thoroughly professional design, but also thoroughly forgettable. The same applies to the interior, neatly set out and with quality fittings, first-class finish, and probably the best factory-fitted navigation systems in the business. But does it leave a lasting impression? The answer is clear.

The Corolla is what millions of people want: the perfect household appliance that does not offend, does not break down—and does not excite. Yet there are murmurings of discontent in Europe, where the conservative Corolla has given way to the more youthful and slightly less middle-of-the-road Auris.

Toyota Hi-CT

Even by the outrageous standards of the way-out concepts that traditionally populate the Tokyo show, this is a strange one. Looking like the front end of a heavy-goods tractor that has mysteriously lost its trailer (or, for those with long memories, the cab of the three-wheeled Scammel prime movers that used to haul goods around British railway yards), the Hi-CT is Toyota's bid to re-ignite the interest of young Japanese buyers whose passions have long since shifted to other types of consumption.

The Hi-CT, says Toyota, offers a new kind of automotive "cool" and "new ways to have fun as a departure from conventional vehicles—an edgy urban vehicle inspired by the thinking and lifestyles of youth." Were it ever to be manufactured, the design would certainly stand out from the crowd—and above it, too. Much taller than it is wide, the Hi-CT is minicar-short, and in fact looks shorter still, as the shallow chassis section, including the rear wheels, sticks out from the main superstructure to leave a useful platform for the transportation of such leisure equipment as bikes or surfboards.

The cab—for that is what best describes it—has proportions never before seen on a car. Very tall and necessitating a step to clamber in, it has deep, sheer sides and a very high waistline; the curved front is undecorated, save for a central Toyota emblem, and the bands of LED headlights are neatly integrated at low level in the front fenders. The rear, surprisingly, has a hatchback-like tailgate, though it is not evident how this could open if equipment were stowed on the rear deck.

The interior is a perhaps predictable feast of strange shapes, unusual materials, and futuristic game-console-like control elements. Next-generation hybrid technology is what will power the Hi-CT, says Toyota, but for most consumers it will be the idea that is a step too far, not the engineering.

Length	3330 mm (131.1 in.)
Width	1695 mm (66.7 in.)
Height	1780 mm (70.1 in.)
Wheelbase	2325 mm (91.5 in.)

Toyota iQ

Design	Toyota European Design Centre
Installation	Front-engined/front-wheel drive
Brakes front/rear	Discs/discs
Length	2985 mm (117.5 in.)
Width	1680 mm (66.1 in.)
Height	1500 mm (59.1 in.)
Wheelbase	2000 mm (78.7 in.)

For many, this tiny car from Toyota was the highlight of both the 2007 Frankfurt show, where it appeared as a near-production concept, and the 2008 Geneva show, where the final version was unveiled. Cleverly named and even more cleverly conceived, the iQ is a small but sophisticated city car that will sit below the Aygo in size but will probably be more expensive to buy, reflecting a growing demographic of consumers willing to pay a premium for advanced design in a small package.

And the iQ is indeed advanced: measuring less than 10 feet from its narrow frontal mouth to the projecting rear wheel arches that form the rear bumper, it manages to sit three adults and a child (or some luggage—but not both) in a footprint little greater than that of the strictly two-seater Smart. This it achieves by shifting the passenger seat further forward than the driver's to allow space behind, while the smaller perch behind the driver is sufficient for a child. Even the suspension and transmission have been rethought in order to free up as much interior space as possible.

But the ingenious internal arrangements are not what is most immediately striking about the iQ; that honor belongs to the intriguing exterior design, which marries soft, compound-curved surfaces with straight-lined themes for a highly distinctive effect. When the car is viewed from the side, the front looks short and stubby, much like that of a Smart; seen from the front, the projecting nose of the concept has been moderated to produce a more homely, rounded effect. The rear, likewise, contrasts the elegant S-shape formed by the wraparound rear window and the rear lights with the straight-lined brow over the rear glass and the straight line at its base.

This is one of the most original and interesting small-car designs of recent years. It could be argued that this is what the multiseater Smart should have been, and Toyota's iQ could prove to be the closest challenger to Smart so far.

Toyota i-Real

The fact that this is the fifth generation of single-seater personal mobility vehicles from Toyota must indicate that the Japanese company is getting serious about such devices. In addition, the choice of "i-Real" as a name could just suggest that this one could even become real. Indeed, Toyota says in its publicity that this concept is a step toward commercialization in the near future.

Following on from the i-Swing and the i-Unit, both from 2005, the i-Real represents yet another change in engineering direction. Whereas the i-Unit ran on four permanent wheels and the more compact i-Swing drove on two wheels but added a third wheel in the front for higher-speed operation, the i-Real has two wheels at the front and a single one at the rear. Like the i-Swing, the i-Real shortens its wheelbase at low speed for extra maneuverability, stretching it again at higher velocities in order to boost stability and handling by lowering the center of gravity.

Toyota has gone to extraordinary technological lengths to ensure that the i-Real can be driven safely in crowded conditions and among pedestrians and cyclists. Sensors scan the entire perimeter of the vehicle to determine the position and speed of surrounding people and objects: if its computer calculates that a collision is imminent, the machine alerts the driver by emitting a noise and vibrating, much like a mobile phone; at the same time, says Toyota, the i-Real alerts people around it of its movements through what the company describes as the "pleasant use of light and sound."

Given that Japan's population is more heavily skewed in favor of older people than those of most other mature economies, there may come a day when the use of such machines as the i-Real is as common as walking. Unfortunately, the typically cautious senior citizen is more likely to be knocked over by a young kid on one of these than to want to try it him- or herself.

Toyota Matrix

Engine	1.8 in-line 4
Power	88 kW (118 bhp) @ 6000 rpm
Torque	156 Nm (115 lb ft) @ 4200 rpm
Gearbox	5-speed manual
Installation	Front-engined/front-wheel drive
Front suspension	MacPherson strut
Rear suspension	Torsion beam
Brakes front/rear	Discs/drums
Front tires	205/55R16
Rear tires	205/55R16
Length	4351 mm (171.3 in.)
Width	1776 mm (69.9 in.)
Height	1539 mm (60.6 in.)
Wheelbase	2601 mm (102.4 in.)
Track front/rear	1514/1496 mm (59.6/58.9 in.)
Curb weight	1225 kg (2700 lb)

Co-developed with the Pontiac Vibe and built at the same joint Toyota–GM plant in California, the second-generation Matrix lends the Pontiac its engineering, its interior, and much of its exterior sheet metal. Yet the two manage to look significantly different, the design's Toyota incarnation being the more cautious and conservative.

In the extensive Corolla range of which it is part, the Matrix is the more practical, family-oriented model; though marketed as a sporty crossover, it is much closer to the European Auris hatchback in format. The four-door Corolla sedan attracts a very different—and rather older—customer.

Compared with the outgoing Matrix the new design is smoother and more coupé-like, in part thanks to the raked wraparound rear window and the deep frontal apron. The bumper has a prominent chin splitting the airflow under the car and into the four separate air inlets. The main central grille is a plain U-shaped aperture, which lacks the personality projected by Pontiac's equivalent—the twin mesh-covered, chrome-rimmed signature grilles.

Much the same holds true for the rest of the design: the blander-detailed solutions chosen for the Matrix make it much less successful than the Vibe. Take the shape of the DLO, for instance. The Toyota's waistline does exactly the same on the front door, kinking upward toward the A-pillar, but while the rear-window shaping of the Matrix is orthodox, the line kicks up sharply to the C-pillar on the Vibe, giving a more dynamic look.

The only area where the Matrix scores visually over the Vibe is at the rear, where the latter's flat tailgate is a real letdown. Overall, however, the pair provide a useful illustration of how small but skillful design touches can transform the look of what is basically the same vehicle. The Toyota looks heavy, blocky, and inert; the Vibe may have equal mass, but it has much more character, poise, and presence.

Toyota RiN

The health and well-being of its occupants is the stated design objective of this Toyota concept, yet another of the weird and wonderful creations to be given the oxygen of global publicity at the 2007 Tokyo auto show. It is a risky proposition to say that this one is crazier than most, but the RiN is certainly a leading contender for a top-five position in the contest.

Looking less like a car and more like a big step-in shower cubicle on wheels, the tall, upright cabin has vertical screens at front and rear and full-depth sliding doors glazed right down to floor level. All the glass is tinted green in order to reduce the amount of ultraviolet and infrared light that reaches the passengers, reinforcing the greenhouse impression; the remaining exterior components—symmetrically shaped at front and rear—are soft white in color and smooth and rounded in design, like domestic fitness equipment. The smoothly curvaceous wheel covers are a good example.

In a neat touch, the headlamps—which automatically adjust to oncoming traffic and even pedestrians—are concealed in slashes in the upper fenders, the same treatment being echoed in green and red at the rear.

Inside the RiN the occupants benefit from a whole array of technologies designed for comfort and serene living. White, beige, and deep green combine to create a calming, harmonious ambience and, says Toyota, "to enrich the complexions of passengers to make them healthier." The seats help the user to maintain a good back posture and are heated for extra luxury; there are an oxygen-level conditioner and a humidifier; and image displays on the panoramic dashboard and in the center of the driver-focused instrument cluster are aligned with the driver's mood.

If urban traffic really is to slow down, congestion to increase, and people to spend even more hours in their cars each day, this madcap idea may prove not to be as mad as it seems.

Length	3250 mm (128 in.)
Width	1690 mm (66.5 in.)
Height	1650 mm (65 in.)
Wheelbase	2350 mm (92.5 in.)

Toyota Sequoia

Engine	4.7 V8 (5.7 V8 also offered)
Power	284 kW (381 bhp) @ 5600 rpm
Torque	544 Nm (401 lb ft) @ 3600 rpm
Gearbox	6-speed automatic
Installation	Front-engined/four-wheel drive
Front suspension	Double wishbone
Rear suspension	Double wishbone
Brakes front/rear	Discs/discs
Front tires	275/55R20
Rear tires	275/55R20
Length	5210 mm (205.1 in.)
Width	2030 mm (79.9 in.)
Height	1895 mm (74.6 in.)
Wheelbase	3099 mm (122 in.)
Curb weight	2722 kg (6000 lb)
Fuel consumption	15.7 l/100 km (15 mpg)

Toyota has gone truly native with the second-generation, US-built Sequoia, making it longer, wider, taller, and more powerful, and adding the option of a 5.7-liter engine to bring it well and truly into the full-size sport utility class.

In effect a people-carrying version of the Tundra pickup, the Sequoia is more than 17 feet long, seats eight, and weighs almost three metric tons. Its design themes are suitably massive, too, with a bold front bumper beam topped by an oversized and heavily chromed rectangular grille extending beyond the headlights and into the hood.

Large running boards and massive 20-inch wheels in generous arches highlight the sheer scale of this vehicle: the huge proportions are doubly evident inside the cabin, where a dual-width center tunnel separates the two front-seat occupants and the broad dashboard has space for an instrument pack containing five circular dials in a line.

Reflecting the exterior styling, the tone of the interior is brash by European standards, though it steers clear of the chrome-clad excesses of some domestic American designs. Characteristically for an American vehicle designed to transport large numbers, the Sequoia provides abundant storage space for all eight passengers' belongings as well as no fewer than eight cup holders. The second row of seats has a sliding adjustment to provide added legroom.

A formidable complement of comfort, safety, and electronic equipment includes triple-zone climate control and even a so-called conversation mirror to allow those in the front to see what is going on in the back seat far behind them.

The Sequoia's style may be big, brash, and insensitive, but it does show that Toyota is just as adept at building an authentic all-American gas-guzzling SUV as it is at producing the world's most popular hybrid vehicle.

Toyota Urban Cruiser

Engine	1.8 in-line 4
Power	95 kW (128 bhp) @ 6000 rpm
Torque	170 Nm (125 lb ft) @ 4400 rpm
Gearbox	5-speed manual
Installation	Front-engined/four-wheel drive
Front suspension	MacPherson strut
Rear suspension	Multi-link
Brakes front/rear	Discs/discs
Front tires	195/60R16
Rear tires	195/60R16
Length	3930 mm (154.7 in.)
Width	1725 mm (67.9 in.)
Height	1540 mm (60.6 in.)
Wheelbase	2461 mm (96.9 in.)
Track front/rear	1486/1491 mm (58.5/58.7 in.)
Curb weight	1191 kg (2625 lb)
Top speed	180 km/h (112 mph)
Fuel consumption	6.5 l/100 km (36.2 mpg)
CO$_2$ emissions	151 g/km

If the Urban Cruiser label sounds familiar, it is because Toyota displayed a concept SUV with that name at the Geneva show in 2006. The 2008 model, far from being a concept, is a full production model.

We would normally praise a company for launching a production model that remained faithful to the style and vision of the concept. But in the case of the Urban Cruiser it is lucky that there is no visible link between the two. The 2006 concept was one of the few truly unfortunate designs to have graced a motor-show podium, its unhappy resemblance to a stranded green frog eliciting a near-unanimous thumbs-down from commentators.

Yet if the new Cruiser gets the thumbs-down, it will be because it is too cautious and uncontroversial. For the design is a straightforward transplant of the US-market Scion xD (reviewed in *The Car Design Yearbook 6*), adapted for European tastes. The Scion, in its turn, is a clone of the Japanese-market Toyota 1st, a slightly chunkier, taller version of the small Yaris hatchback. However, in becoming the Urban Cruiser, the design has lost the odd, block-like front grille and bumper of the Scion, and has reverted to a standard corporate Toyota nose, less distinctive even than the European-styled Yaris or Auris.

The result is a small, high-sided vehicle that is neat and pleasant rather than overtly attractive; it has a nice stance in the road, thanks to good wheels and tires and well-defined wheel arches; and the thick C-pillar and the slight upturn of the roof to form a spoiler lip over the rear window are recognition features. The headlights, set slightly back from the stepped, two-layer grille, form the beginning of a feature line that runs back along the shoulder to the tops of the rear lights. Blacked-out front and rear apron panels hint at off-road adventure. But if Toyota believes Urban Cruiser will equate to urban cool, it is in for a disappointment.

Toyota Venza

Engine	3.5 V6 (2.7 in-line 4 also offered)
Power	200 kW (268 bhp) @ 6200 rpm
Torque	334 Nm (246 lb ft) @ 4700 rpm
Gearbox	6-speed automatic
Installation	Front-engined/all-wheel drive
Brakes front/rear	Discs/discs
Front tires	245/50R20
Rear tires	245/50R20
Length	4800 mm (189 in.)
Width	1905 mm (75 in.)
Height	1610 mm (63.4 in.)
Wheelbase	2776 mm (109.3 in.)

Designed in California, engineered in Michigan, and to be built in Kentucky, the Toyota Venza is shaped around American tastes in order to sell to Americans.

The Venza is what the car business labels a segment-buster, seeking to combine key qualities from different vehicle segments (in this case, the comfort of sedan cars and the versatility of an SUV) in order to attract a greater number of customers to a single product. If one looks at the long sweeping arc of the Venza's passenger compartment, its muscular nose, and its slanted hatchback rear end, it would appear that Toyota has been listening to the same market research that led Mercedes-Benz to introduce the R-Class in 2005. The R-Class also sought to blend SUV, sedan, and station-wagon qualities; unfortunately, it has failed to take off in the market, mainly because buyers are not sure what it is or what it can do.

Toyota, several price rungs further down the market, is not likely to make the same mistake: the Venza is more conservatively styled than either the Mazda CX-7 or the Nissan Murano, likely competitors approaching the same customers from the SUV angle, and by downplaying any 4x4 or crossover SUV design cues it advertises itself more strongly as a comfort-oriented family five-seater.

The Venza is a good deal more conservative than the bold 2005 FT-SX concept that was its inspiration; even so, it has a discreet elegance to its profile, especially the strong D-pillar, which recalls the earlier Lexus IS Sportback. Viewed from the rear, the complex taillights give lots of visual energy, but there is an awkward area above the number plate where a spoiler lip is forcibly drawn out below the tailgate glass.

Inside, the conservative Toyota interior is unrecognizable from the innovative FT-SX concept, instead combining orthodox materials in a conventional manner, the only unusual feature being the gearshift located on the left-hand edge of the center stack.

Volkswagen Passat CC

Engine	3.6 V6 (2.0 in-line 4, and 2.0 diesel, also offered)
Power	221 kW (297 bhp) @ 6600 rpm
Torque	350 Nm (258 lb ft)
Gearbox	6-speed automatic
Installation	Front-engined/all-wheel drive
Front suspension	MacPherson strut
Rear suspension	Multi-link
Brakes front/rear	Discs/discs
Length	4796 mm (188.8 in.)
Width	1856 mm (73.1 in.)
Height	1420 mm (55.9 in.)
Track front/rear	1553/1557 mm (61.1/61.3 in.)
0–100 km/h (62 mph)	5.6 sec
Top speed	250 km/h (155 mph) limited

It is perhaps more constructive to see the Passat CC not, as many commentators have, as a reduced-scale model of the Mercedes-Benz CLS, but as a handy way of edging the Volkswagen brand upmarket, without too much investment and without the indignity of openly sacrificing the slow-selling Phaeton.

For sure, the parallels with the CLS are tempting: some of the personalities involved were the same; the overall silhouettes are similar, the DLO profile especially so; the trunk lid continues the same slope as the rear window. The list goes on.

The CLS has been a big success; Volkswagen clearly has the same hopes for the CC, though perhaps for different reasons. Whereas the CLS is simply a sportier-looking alternative pitched between the existing E- and S-Class lines, the Passat CC is effectively on its own at the top of VW's mainstream sedan range. More importantly, it will allow the VW buyer to be slightly indulgent, slightly individual in his or her choice, making it an attractive alternative to the depressingly sober-sided standard Passats sold by the hundreds of thousands each year. Such individuality has its price, however: as with most coupés, the buyer pays more—several thousand more—but what is gained in style is lost in practicality.

Compared with the dull Passat sedan the CC is indeed stylish; arguably, it is a more successful result than the Mercedes, the sweeping roofline giving the currently fashionable four-door coupé look with relatively little sacrifice in headroom for the two rear occupants, who have their own individual seats.

The rest of the interior remains disappointingly unchanged from the standard car, save for a new steering wheel and such upmarket options as ventilated front seats. A shrewd marketing exercise, the Passat CC has responded quickly in bringing the big-car coupé vogue down into the more affordable middle-class sector; critically, it has found the flair that is missing from the standard Passat.

Volkswagen Routan

Design	Chrysler, Klaus Bischoff
Engine	4.0 V6 (3.8 V6 also offered)
Power	188 kW (253 bhp)
Torque	357 Nm (263 lb ft)
Gearbox	6-speed automatic
Installation	Front-engined/front-wheel drive
Front suspension	MacPherson strut
Rear suspension	Torsion beam axle
Brakes front/rear	Discs/discs
Front tires	225/65R16
Rear tires	225/65R16
Length	5140 mm (202.4 in.)
Width	1950 mm (76.8 in.)
Height	1750 mm (68.9 in.)
Wheelbase	3080 mm (121.3 in.)
Track front/rear	1650/1645 mm (65/64.8 in.)
Top speed	200 km/h (125 mph)
Fuel consumption	14.7 l/100 km (16 mpg)

Visitors to the 2008 Chicago show could have been forgiven for a sense of *déjà vu* as they strayed on to the Volkswagen stand. For there, in among the Jettas, the Passats, and the new Beetles, was a large black minivan, labeled Routan and looking uncomfortably big and boxy alongside its smoothly styled German brand-mates.

Study the Routan more closely and further familiar features come to light: six deep, flat windows, twin sliding rear doors, a broad rear tailgate. Truth to tell, the Routan is no genuine VW but a made-over Chrysler, a Voyager topped and tailed with VW styling cues to give the German company a much-needed large minivan for the North American market. And, on the whole, Volkswagen's designers have done a reasonable job of disguising the origins of the vehicle.

While the sides have been kept simple and, in the Volkswagen idiom, free from adornment, there was little VW could do to hide the slabby flanks of the donor model; the paired solid-chrome door handles are a classy touch, however. The makeover has been more extensive at the rear and, especially, at the front. In place of Chrysler's boxy, truck-like hood comes a smoother, more sweeping, clamshell affair, fronted by a discreet VW grille and flanked by much larger headlamps drawn backward over the wheel arches and characterized by VW's familiar wavy line at their bases. At the rear, likewise, a passably Volkswagen look has been created thanks to larger, reshaped taillights, again with curving baselines.

Inside the voluminous six-plus-seater cabin Volkswagen has again revised and refined the Chrysler original. The materials are of a higher grade, the steering wheel carries the VW logo and, though the odd vertical transmission shifter is retained, the ugly and dominant full-height center stack of the Chrysler is softened in favor of a more attractive and less intimidating conventional design.

Volkswagen Scirocco

Engine	2.0 in-line 4 (1.4, and 2.0 diesel, also offered)
Power	147 kW (197 bhp) @ 6000 rpm
Torque	280 Nm (206 lb ft) from 1700 rpm
Gearbox	6-speed DCT
Installation	Front-engined/front-wheel drive
Front suspension	MacPherson strut
Rear suspension	Multi-link axle
Brakes front/rear	Discs/discs
Front tires	225/45R18
Rear tires	225/45R18
Length	4256 mm (167.6 in.)
Width	1810 mm (71.3 in.)
Height	1400 mm (55.1 in.)
Wheelbase	2578 mm (101.5 in.)
Track front/rear	1569/1575 mm (61.8/62 in.)
Curb weight	1300 kg (2866 lb)
0–100 km/h (62 mph)	7.2 sec
Top speed	235 km/h (145 mph)
Fuel consumption	7.7 l/100 km (30.5 mpg)

Scirocco is a highly emotive word among Volkswagen aficionados and is often equated with an era when the German company produced what fans regard as real sports cars. So the return of the Scirocco nameplate on a brand-new sports coupé, presented at the 2008 Geneva show, was bound to be a very emotional affair, with customer expectations riding sky-high and industry commentators eager to see Volkswagen revealing its long-awaited fun side.

Eighteen months before, VW previewed the Scirocco with the Iroc concept, combining a dramatic frontal style with a practical four-seater coupé body. The production car keeps the long, low style of the Iroc, but softens it with a less controversial front end and a more restrained treatment at the rear.

Instead of the bold, six-sided air-catcher that dominated the front of the Iroc, the Scirocco sports a narrow slot grille above the bumper, flanked by long lights that extend outward rather than upward, to emphasize the car's width. The smooth, crowned hood flows into a suitably fast windshield.

At the side, the standout features include the long, drawn-out DLO with the rising rear-window line that references the original 1974 car, and the pinched-in sides that catch the light and highlight the car's leanness and length. A further key characteristic is the way the cabin tapers in plan toward the rear, leaving substantial shoulders over the rear wheel arches. This gives a powerful, though not particularly sporty, impression from behind.

Inside, too, the ambience is Golf-like rather than overtly sporty. For a coupé it has good legroom in the back, though passengers may find the narrow windows claustrophobic; there is some neat detailing in the fixtures and fittings, but little originality.

Overall, designer opinion appeared to be mixed about the new Scirocco. It is elegant rather than explicitly dynamic, and the need for practicality has made it bulkier than its forebear. But seen against its likely competitors—the Audi A3, the Mercedes CLC, and the Volvo C30—it appears to check off all the right boxes.

Volkswagen Tiguan

Engine	2.0 in-line 4
Power	147 kW (197 bhp)
Torque	280 Nm (206 lb ft) @ 1700–5000 rpm
Gearbox	6-speed manual
Installation	Front-engined/all-wheel drive
Front suspension	MacPherson strut
Rear suspension	Multi-link
Brakes front/rear	Discs/discs
Front tires	235/55R17
Rear tires	235/55R17
Length	4430 mm (174.4 in.)
Width	1810 mm (71.3 in.)
Height	1670 mm (65.7 in.)
0–100 km/h (62 mph)	7.6 sec
Top speed	211 km/h (131 mph)
Fuel consumption	9.1 l/100 km (25.8 mpg)

This car was shown in near-production form as the Tiguan concept at the 2006 Los Angeles show. The few changes that have been made to the vehicle in the intervening period have served principally to tame down the design even further—to the point where it can be guaranteed not to offend the ultraconservative American buyer. The contrast with the vigorous and aggressive Ford Kuga, which occupies exactly the same medium car–SUV crossover ground but which will not be going to the USA, could not be greater: this shows just what a depressing effect North American consumer focus groups can have on originality and spark in vehicle design.

The Tiguan comes across in most senses as a scaled-down version of VW's large luxury SUV, the Touareg—no bad thing in view of the success and prestige of that model. A shallow clamshell hood sits on top of the grille and headlights, its base at the rear forming the baseline for the side windows; the restrained surfacing and sober lines give a feeling of reassured confidence but not a lot of passion.

A raised, body-color moulding along the sides of the vehicle, aided by a broad blacked-out sill section placed well above the axle center lines, helps to reduce the visual mass of the body side. The rear, with its smooth, elliptical lights and gentle curves, is perhaps the most attractive aspect of the design.

One interesting point about the Tiguan is the availability of two different versions aimed at different buyer groups. The standard model has a longer and deeper frontal structure and greater overhang at the front, features normally frowned upon by serious off-road enthusiasts as they limit the machine's ability to approach steep climbs. These users are catered to by the second version, with its more compact front understructure allowing greater off-road ability.

Volkswagen Up! & Space Up!

The VW Up! is not one but a whole family of new-generation cars exploring variations on a common theme: that of a compact city-commuter. All (there are three: the Up!, the Space Up! and the Space Up! Blue) share a distinct design language that is pure and minimalist, with simple surfacing, generally flat panels, and restrained detailing. Despite their flat sides, they avoid looking vanlike thanks to subtly swollen wheel arches flowing into contoured sills, and all three introduce a new design idea that could catch on: taillights mounted inside, visible through the smoked-glass tailgate.

The basic three-door Up! (white) looks neat and sporty, despite measuring just about 138 inches in length; it has a faster windshield and an upward kink in the window line at the C-pillar. The longer five-door Space Up! (blue) looks more formal and upright, while the LA show's Space Up! Blue, paradoxically finished in pale silver-green, goes one better, with a raised roofline and classic VW Microbus-style extra windows set into the edges of the roof, and a large panel of solar cells occupying the center of the roof.

All share similar frontal themes: a big central VW logo linked via a narrow slit to an elongated headlight at each side; the bumper bar floats in a black surround (echoing the style at the rear); and there is no need for a grille thanks to the rear mounting of the engine. Aside from stating that the LA concept had a high-temperature fuel-cell powerplant, VW has said little about the diesel and gas engines in the other two versions.

Inside themes are similar, too, with a very simple, slim-looking dashboard with minimal instrumentation and few visible controls; the seats are thin in order to maximize interior space, but have a series of separate sections that inflate to provide the required support.

The Up! series represents welcome fresh thinking from Volkswagen; the task now is to match this impressive blend of function, form, and fun with the economic realities of the production line.

Details refer to the Space Up! model

Design	Walter de' Silva
Installation	Rear-mounted/transverse/rear-wheel drive
Front tires	165/50R18
Rear tires	165/50R18
Length	3680 mm (144.9 in.)
Width	1630 mm (64.2 in.)
Height	1540 mm (60.6 in.)
Wheelbase	2560 mm (100.8 in.)
Track front/rear	1420/1420 mm (55.9/55.9 in.)

Volvo XC60

Design	Steve Mattin
Engine	3.0 in-line 6 (2.4 in-line 5 diesel also offered)
Power	210 kW (282 bhp) @ 5600 rpm
Torque	400 Nm (295 lb ft) @ 1500–4800 rpm
Gearbox	6-speed automatic
Installation	Front-engined/four-wheel drive
Front suspension	MacPherson strut
Rear suspension	Independent coil springs
Brakes front/rear	Discs/discs
Front tires	235/60R18
Rear tires	235/60R18
Length	4628 mm (182.2 in.)
Width	1891 mm (74.5 in.)
Height	1672 mm (65.8 in.)
Wheelbase	2774 mm (109.2 in.)
Track front/rear	1632/1586 mm (64.3/62.4 in.)
Curb weight	1990 kg (4387 lb)
0–100 km/h (62 mph)	7.5 sec
Top speed	210 km/h (130 mph)

There was near-universal praise for new Volvo design director Steve Mattin's execution of the XC60, the Swedish company's long-awaited competitor to the likes of the Land Rover Freelander, the BMW X3 and, soon, the Mercedes-Benz GLK, on the first rung of the premium SUV market.

The XC60 is clearly identifiable as a close relation of the larger XC90 and, just as with the bigger vehicle, Volvo has acknowledged the social and environmental concerns surrounding SUVs and has been skillful in taking any aggression out of the car's looks and allowing it to project a more family-friendly image. In particular, the front end is softer and more rounded than in most competitors, and the wheels, though big, are not highlighted by exaggerated swollen arches.

Overall, the XC60 is more sculpted, and hence more energetic, than the XC90. As with the larger car, a prominent rounded shoulder line runs from the hood through to the taillamps. In this instance, however, there is an undercut depression below this feature line, drawing the sides in and reducing the apparent bulk of the vehicle.

The taillamps offer a striking graphical aesthetic at the rear, accentuating not only the curvature of the body side but also the dramatic forward slant of the C-pillar. Between the rear lights, the tailgate and rear-window glass are relatively plain, thus drawing the eye outward. All round the vehicle, dark-colored sills and aprons again serve to reduce the visual height and mass of the body. At the front, the optional aluminum rock-shield is cleverly extended up to form the lower grille surround and the number-plate mounting, while the reshaped main grille, with its enlarged Volvo emblem and vertical outboard running lights, is expected to become the template for future Volvo designs.

Inside, the XC60 naturally comes up to Volvo's familiar high standards, and the vehicle is also notable as the first to feature the company's pioneering City Safety system (see Vehicle Sensing Technologies feature, page 18), designed to minimize low-speed collisions.

Zagato Bentley GTZ

Design	Norihiko Harada at Zagato
Engine	6.0 W12
Power	447 kW (600 bhp)
Torque	750 Nm (553 lb ft)
Gearbox	6-speed automatic
Installation	Front-engined/all-wheel drive
Front suspension	Double wishbone
Rear suspension	Multi-link
Brakes front/rear	Discs/discs
Front tires	275/40R19
Rear tires	275/40R19
Length	4804 mm (189.1 in.)
Width	1965 mm (77.4 in.)
Height	1390 mm (54.7 in.)
Wheelbase	2745 mm (108.1 in.)
Track front/rear	1623/1607 mm (63.9/63.3 in.)
Curb weight	2350 kg (5181 lb)
0–100 km/h (62 mph)	4.6 sec
Top speed	319 km/h (198 mph)
Fuel consumption	16.6 l/100 km (14.2 mpg)
CO$_2$ emissions	396 g/km

Ever since the demise of the old-fashioned prewar coachbuilders and their often extravagant designs on high-class chassis, there has been something of an unspoken taboo when it comes to reinterpreting the style of such truly noble makes as Rolls-Royce, Bentley, and even Aston Martin. It is territory where outside designers generally fear to tread. But that has not deterred Italian carrozzeria Zagato, famed for its offbeat and sometimes hard-to-fathom designs, from turning its hand to the Bentley Continental GT.

The result, the Zagato GTZ, sat on a plinth uncomfortably close to Bentley's 2008 Geneva show stand and drew a mixture of responses from Bentley personnel and other onlookers. Some welcomed its free interpretation of Bentley styling cues and the classic Bentley grand-touring ethos; others looked askance, uncertain whether it counted as sacrilege or mischief.

Zagato's main achievement is to have made the GTZ more voluptuous, by adding more flowing forms to its surfacing, notably the pronounced rear hips, which recall the immediate postwar R-Type model. The rear, in particular, is much more flamboyant, with a smooth, boatlike tail, elegant elongated vertical taillights on the rear fenders, and a steeply raked rear windshield leading into the distinctive double-bubble roofline.

From the front, the Continental cues are more clearly preserved, though again the stance is more sporting, with a bigger, deeper Bentley mesh grille, a wide lower air intake, and perhaps superfluous air exit slats puncturing the hood. The Bentley four-headlamp layout is retained, but Zagato has added indicator side lights neatly faired into the front edges of the front fenders.

Inside, Zagato has attempted the near-impossible task of making a Bentley even more sumptuous than it already is, with a reshaped center console among the changes. Bentley purists may question the ethics of putting an angle-grinder to a Continental, but in reality such work as Zagato's is the continuation of a tradition of aesthetic craftsmanship and originality—and should be welcomed.

Zagato Bentley GTZ **Production** 287

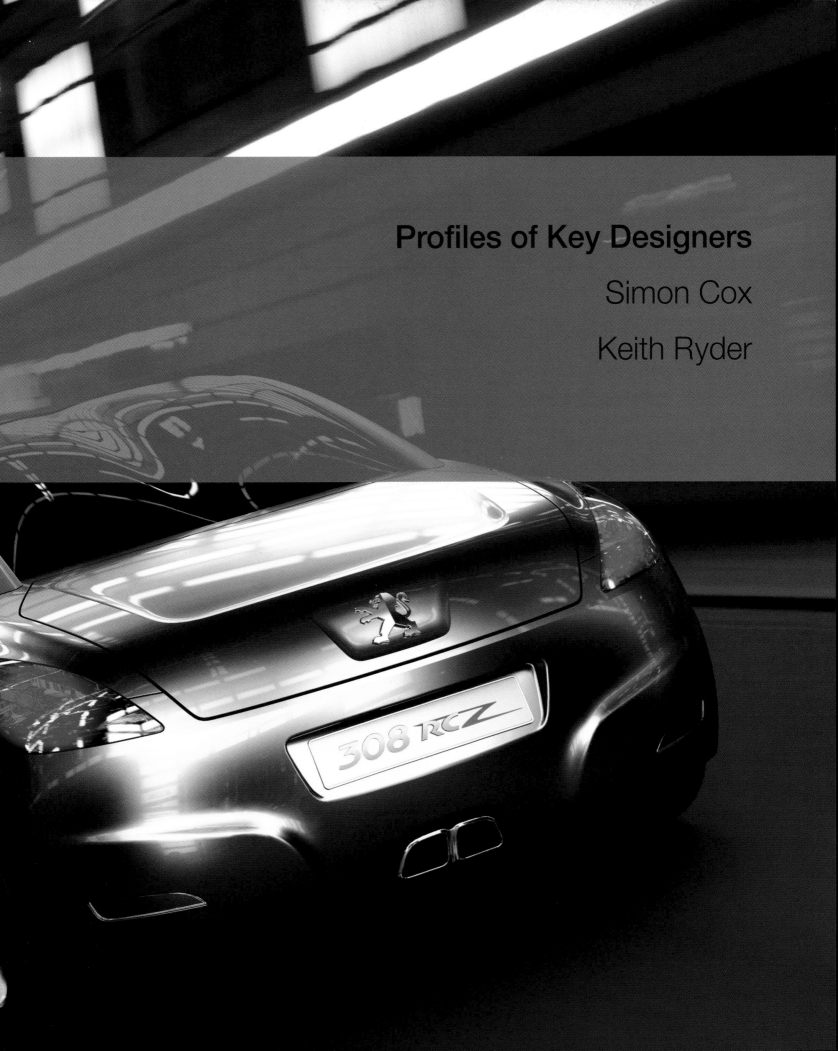

Profiles of Key Designers

Simon Cox

Keith Ryder

Simon Cox

Today's Cadillacs are a far cry from the flashy, finned monsters of US movie legend. But with their sharply chiseled features, intricate angles, and big on-road presence, they are no less impressive (and no less American) in their style.

So it comes as something of a surprise to learn that the inspiration behind those archetypal Stateside premium products is not a squad of Americans in a huge studio in downtown Detroit but a quiet Englishman operating—in relative isolation—from an atelier in the British Midlands.

Simon Cox has been running General Motors' Advanced Design Studio in the UK for a decade, during which time he has produced a series of radical and sometimes startling concepts that helped to turn Cadillac from something of an aesthetic embarrassment in the automotive world into a fresh, modern nameplate with one of the clearest brand identities of any premium vehicle maker.

Cox, who was born in Warwickshire, not far from where he now works, has in fact been in GM's orbit for much of his professional working

life. Prior to his current position he was chief designer at Isuzu, a Japanese carmaker then controlled by GM, while between 1986 and 1990 he was senior designer at Lotus, at that time also controlled by GM.

Cox's first job in the car business, however, was as a sculptor in advanced design for Peugeot-Citroën, where he helped shape the Citroën Eco 2000 concept—a highly advanced design for its era. Sculpture came naturally to the young Cox, for his original training was in three-dimensional design, jewelry and silversmithing, followed up by a master's degree in industrial design engineering. But by this stage, having worked for Peugeot for two years and seeing inspiration in the sculptural forms of the Porsche 928, the car bug was really beginning to bite. With backing from Ford, Cox gained a highly prized place on the famous postgraduate automotive design course at London's Royal College of Art.

Upon graduation Cox moved straight to Lotus, rising to senior designer and generating

Top
The Cien concept of 2002 is cited by Simon Cox as his biggest professional challenge: it had to convince GM that there were no limits to Cadillac's potential.

Opposite, top
The Porsche 928: an inspiration for Cox.

Opposite, bottom
The CTS sedan confirmed Cox's distinctive angular design language as a selling point for Cadillac.

a wide variety of designs for both Lotus and external clients. The latter included Triumph motorcycles—Cox confesses to a lifelong fascination with bikes, especially Ducatis and Tamburini's MV Agusta F4—as well as Dodge, Chrysler, Lamborghini, and Isuzu concepts, notably the much-praised Isuzu 4200R. For Lotus, Cox worked on such programs as the Esprit and the short-lived front-wheel-drive Elan.

Moving to Isuzu in 1990, Cox gained steadily greater responsibilities through further concept models such as the Como and the Deseo. The VX2 4x4, which he lists as the first program where he had overall design responsibility, was a big hit at the 1997 Tokyo show and went on to win Best Concept at the Detroit show two months later. While at Isuzu,

Cox also designed a vehicle that would become a familiar sight on European roads: the Opel/Vauxhall Frontera, GM's first venture into the 4x4 market and based on Isuzu architecture. Prior to this, Cox had scored an unexpected hit with the segment-busting Isuzu Vehicross, which can be seen as the forerunner of today's generation of leisure-oriented crossover SUVs.

With the shift to GM's Advanced Design Studio in 1998 came the challenge to move the image of GM's North American brands into the new century. This, as Cox concedes, was a considerable task; indeed, he cites the Cadillac Cien concept of 2002 as the toughest professional challenge of his career to date.

"The big challenge was convincing GM that the Cadillac brand had no limits in terms of future

product," Cox recalls. Indeed, the Cien—Spanish for "hundred," and commissioned to celebrate GM's centenary—was a low-slung, mid-engined, ultra-high-performance super sports car powered by a V12 engine and embodying all the latest technologies. A greater contrast to Cadillac's then-current big, lazy sedans and lumbering SUVs could hardly be imagined.

"The timing and complexity of the project were a challenge too," continues Cox. "It was a fully working concept, with a bespoke carbon-fibre monocoque chassis, a specially developed V12 engine, and a unique transmission and suspension system. It was in many respects a real vehicle, designed and built in a short time."

Fittingly, the Cien again won Best Concept at Detroit for Cox, and, although it was never put

Opposite
The Cadillac Imaj of 2000 was Cox's proposal for a large four-seater luxury sedan; its specification included night vision and head-up display systems.

Above
The larger-than-life Sixteen concept won awards in 2003; its role was to explore ideas for ultraluxury Cadillacs, though this course was later abandoned.

Right
A racing version of the Chevrolet Corvette: several of Cox's suggestions were included in the design of the production model.

Above
The GMC Graphyte, shown at Detroit in 2005, proposed a softer design language for SUVs, both inside and outside; the powertrain incorporated a hybrid element.

Right
The Opel GT roadster was inspired by Cox's design for the Vauxhall Lightning concept car, built to celebrate Vauxhall's centenary in 2003.

into production, its design language of sharp surfaces, angled cuts, and ground-hugging stance is to be seen on every model in the current Cadillac catalog.

The following year, 2003, Cox put down a further marker for Cadillac's product renaissance: the imposing, not to say gigantic, Sixteen concept. With its massive wheels, its lengthy hood and its huge 16-cylinder engine, it was a modern take on the luxury limousine of the prewar golden age. And, although it, too, failed to make production, it sent out a signal that Cadillac was eager to compete in any segment, no matter how high.

From his UK base Cox has had a hand in many US production cars too, notably the well-received Saturn Vue. He also delivered the Vauxhall Lightning concept, which influenced the later GM Solstice sports cars.

Asked to name the inspirations for his work, Cox cites a list that is artistic and architectural as well as automotive. "My inspiration was varied," he says. "It came from such artists as Leonardo da Vinci, Henry Moore, Le Corbusier, and Frank Lloyd Wright, to Ferdinand Porsche, Enzo Ferrari, and Sergio Pininfarina."

Unquestionably, Cox has been one of the leading lights in bringing Cadillac its new look and its renewed favor with customers; unquestionably, too, he has a great deal more to give—both at Cadillac and at the other GM brands, where his outsider's perspective, often controversial, often provocative, is said to be greatly valued by those inside the GM hothouse.

Keith Ryder

It was a high-speed ride in a glamorous sports car that set the young Keith Ryder on course to becoming a car designer. "At the age of about fifteen, I was driven at speeds of up to 100 mph in an Austin Healey 3000," he recalls. "The sound of the engine and the superb looks were so awe-inspiring." That Healey, says Ryder, is what impelled him to become a car designer.

Ryder's first steps on leaving school took him to Hartlepool College of Art in northeastern England, after which he obtained a three-year Bachelor of Design degree in industrial design engineering at Leicester Polytechnic. From there, Ryder embarked on the path trodden by so many famous designers, both before and since, enrolling in the highly respected postgraduate automotive design course at London's Royal College of Art.

Graduating from the RCA in 1982, Ryder was one of a hugely influential wave of British car designers who fanned out across most of the global automobile industry, and many of whom are in key positions at leading carmakers today.

For Ryder the die was cast pretty soon after graduation. Though he was sponsored through the course by Austin Rover, he was soon snapped up by Talbot's advanced design studio in Whitley, near Coventry. The studio was renowned during the 1980s as an incubator for top talent in automotive design. When Peugeot, which had taken over the bankrupt Chrysler Europe and rebranded its models Talbot, announced that it intended to close the studio, Ryder jumped at the offer of a move to the Centre Style Peugeot near Paris. Now, a quarter of a century and many scores of designs later, he is in charge of design for a wide range of Peugeot models and, as the effective number two in the system, reports directly to Jérome Gallix, overall head of Peugeot design.

It was while he was still at the Talbot studio in the UK that Ryder worked on his first major design—the ambitious Eco 2000 research concept, co-funded by the French government, Peugeot-Citroën, and Renault. Interestingly, the other designer profiled in this edition of *The*

Top
Keith Ryder has overseen the complete Peugeot 207 lineup, one of Europe's top-selling ranges.

Opposite, top
The glamorous Austin Healey 3000, the car that inspired Ryder to become a car designer.

Opposite, bottom
Ryder's Peugeot 307 design (coupé-cabriolet shown) combines a high architecture with stylish looks.

Car Design Yearbook, Simon Cox, was also on the same project as the young Ryder, working as a clay sculptor.

During his initial years at the Centre Style Peugeot, Ryder's work was concentrated on interior design for both production and prototype cars; the first design of his to reach the showroom was the restyle of the fascia and interior of the Peugeot 205, begun in 1984. This was already a major responsibility: the 205 had been a spectacular success, saving the whole Peugeot empire from near-meltdown in the mid-1980s, and the scrappy interior of the 1983 original had been the only real point of criticism. (The exterior of the 205, incidentally, was never

facelifted despite a fifteen-year production life.)

Soon afterward the smaller Peugeot 106 appeared, again with a Ryder-designed interior, and was met with almost as much acclaim. Ryder went on to design and project-lead the interior for the 406 sedan, a key model that aimed to present a classy, upmarket feel in a volume production car, something that had until that point eluded both Peugeot and its sister brand Citroën.

In parallel with the work on the 406, Ryder was able to extend his vision significantly with the Oxia concept car. Presented at the Paris motor show in 1988, the Oxia was a glamorous mid-engined sports car that set out to lift Peugeot's image above that of a generalist

manufacturer. Ryder's interior was notable for its smooth, clean planar surfaces, restful colors, and clear graphics. "The Oxia was our first running prototype supercar," recalls Ryder. "Along with the 205 GTI it created a great impact that placed our young design team on the map."

Greater responsibilities were to follow. Peugeot's strategy of replacing the all-conquering 205 with two models—the smaller 106 and the larger 306—had not enjoyed the success expected of it. Though Peugeot managers continued to insist that combined 106 and 306 sales were greater than the 205's had been, the split strategy had backfired in one very serious way: the abandonment of the 205 line had

Opposite and left
The 307 Cameleo concept of 2001 explored a combination of sedan, hatchback, and pickup design ideas and acted as a teaser for the 307 SW launched later the same year. The interior combined novel textures and finishes, paving the way for the interchangeable colors in the 1007 model in 2004.

Above
The final design of the Peugeot 308 hatchback, chosen from several proposals, extends the design language of the highly successful 207.

opened the door to what would prove to be a deadly rival—the Renault Clio. Chic, smart, and also practical, the Clio quickly captured the hearts of those who had previously been driving 205s. Peugeot knew that unless it won these buyers back quickly, it would lose them for good.

The result was the 206, which came out in 1998 but which had been five years in the making. Ryder had been appointed design manager for Platform 2, the new supermini-sized architecture that would figure strongly in PSA–Peugeot Citroën's much more rational future platform strategy. The 206 was to be the first car on this platform, and Ryder again led the interior design work.

The big success of the 206 proved that Peugeot and its designers had not lost their touch; Ryder, meanwhile, was busy planning a much bigger push in the middle market, where the elegant 306 was beginning to show its age technically, if not aesthetically. Given overall design responsibility for the 307, Ryder cites it as one of the biggest challenges of his career: "It was an innovative design, with a cab-forward concept and tall packaging. This was a real challenge for both engineers and designers."

The semi-high hatchback 307 hit the streets in 2001, and it was voted European Car of the Year for 2002 by leading European motoring journalists. However, this was not the only shape

in the 307 family: in succeeding years, Peugeot introduced the idea of the SW—a high-class wagon with an extensive glazed roof—as well as the CC coupé convertible, and each of these production models was trailed by a motor-show teaser concept. There was even a sedan model, aimed at the Chinese market.

After steering the advanced design phases of both the interior and the exterior of the 207, the success of which was vital to PSA's fortunes, Ryder—by now design director for midsize vehicles – moved on to the equally ambitious 308 program. The 207 had popularized the now-familiar broad gaping mouth frontal identity for the Peugeot brand: the 308 took these looks and

scaled them up to the family hatchback size class. Yet, building on the basis of the 307, the 308 was also designed to take the tall car concept to the next stage with what Ryder calls "a new balance of exterior volumes and exterior styling." This, he says, is the design of which he is most proud. The interior has a coupé-type central console that is steeply raked. "This provides a great style base for the interior," he states, "but it was a big challenge for the engineers."

Again, the 308 is a big family of models, covering the same bases as its predecessor, but with one possible addition. For the 2007 Frankfurt auto show Ryder and his colleagues came up with a 308-based concept that pushed the Peugeot's

design in a more glamorous direction. Low, wide, and curvaceous, the 308 RC Z coupé prompted strong reactions at its unveiling, not least for its unusual proportions, with its long trunk, steeply raked rear window to the abbreviated cabin, and a provocatively cab-forward stance in profile.

Some likened the RC Z to a modern-day Karmann-Ghia VW; others described it as Peugeot's equivalent of the Audi TT. One way or another, the debate stirred up showed a more adventurous Peugeot, a Peugeot once again prepared to mix it up with the premium brands when it comes to style—and substance. And for this, Keith Ryder must justly take a big share of the credit.

Opposite
The racy 308 RC Z coupé polarized opinions when it was unveiled in 2007 but is now set for production as an image leader for a revitalized Peugeot brand.

Above
The Peugeot Promethee concept of 2000 previewed the theme and many of the ideas seen on the production 307 SW a year later, including the windshield that wraps over into the roof. The asymmetric door layout was dropped for the showroom model.

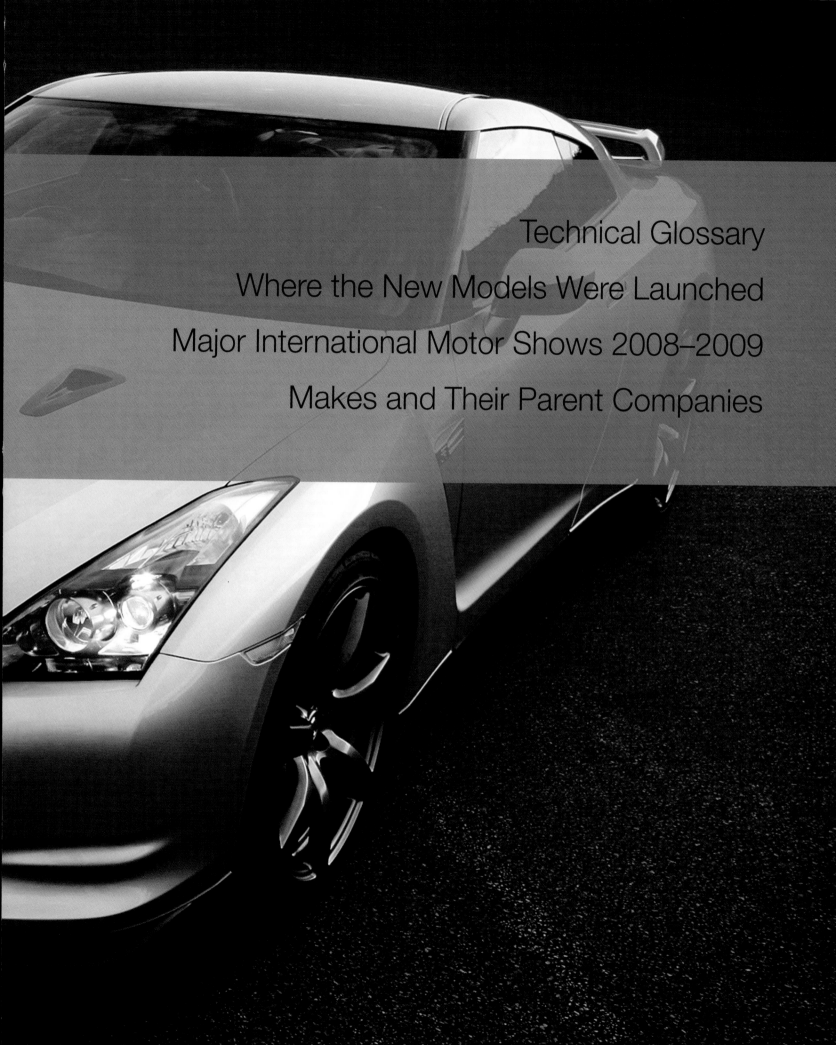

Technical Glossary

Where the New Models Were Launched

Major International Motor Shows 2008–2009

Makes and Their Parent Companies

Technical Glossary

Specification tables

The following list explains the terminology used in the specification tables that accompany the model descriptions. The amount of data available for any given model depends on its status as a concept or a production car. More information is usually available for models currently in or nearing production.

engine	Engine size is quoted in liters, and refers to the swept volume of the cylinders per crankshaft rotation; 6.0, for example, means a 6-liter (or 6000-cc) engine. "In-line" or "V" followed by a number refers to the engine's number of cylinders. An in-line 4 engine has four cylinders in a single row, while a V8 engine has eight cylinders arranged in a V-formation. A flat-four engine has four cylinders lying in a horizontal plane, two opposing each other. In-line engines of more than six cylinders are rare today because they take up too much packaging space; an in-line 12, for instance, would require a very long hood. Only Volkswagen makes a W12, an engine with its twelve cylinders arranged in a W-formation. The configuration of cylinders is usually chosen on cost grounds: the higher the car's retail price, the more cylinders product planners can include.
power	Engine power is given in both metric kilowatts (kW) and imperial brake horsepower (bhp). Both are calculated at optimum engine crankshaft speed, given in revolutions per minute (rpm) by manufacturers as a "net" measurement—in other words, an engine's output after power has been sapped by other equipment and the exhaust system—and measured by a "brake" applied to the driveshaft.
torque	Simply the motion of twisting or turning, in car terms torque means pulling power, generated by twisting force from the engine crankshaft. It is given in Newton meters (Nm) and pounds feet (lb ft). The higher the torque, the more force the engine can apply to the driven wheels.
gearbox (transmission)	The means by which the power is transmitted from the engine or motor to the road wheels. On cars with gasoline or diesel engines this takes the form of either a manual gearbox and clutch, or a fully automatic system. More recently, carmakers have begun offering automated manuals, where the clutch is actuated automatically and gearshifts can be programmed or triggered manually in sequence. A refinement gaining popularity is the dual clutch transmission (DCT), which offers the smoothness of an automatic but even better economy than a manual. Another automatic system is the CVT, or continuously variable transmission, where the ratio changes steplessly. Electric cars have motors that produce strong torque (see above) as soon as the motor starts turning; they do not generally need either a clutch or a gearbox, though as speeds and power rise some are now appearing with automated two-speed gearboxes to enable a higher top speed.
suspension	All suspension systems cushion the car against road or terrain conditions to maximize comfort, safety, and roadholding. Heavy and off-road vehicles use "rigid axles" at the rear or front and rear; these are suspended using robust, leaf-type springs and steel "wishbones" with "trailing arms." "Semi-rigid axles" are often found at the back on front-wheel-drive cars, in conjunction with a "torsion-beam" trailing-arm axle. "Independent" suspension means each wheel can move up and down on its own, often with the help of "trailing arms" or "semi-trailing arms." A "MacPherson strut," named after its inventor, a Ford engineer called Earl MacPherson, is a suspension upright, fixed to the car's structure above the top of the tire. It carries the wheel hub at the bottom and incorporates a hydraulic damper. It activates a coil spring and, when fitted at the front, turns with the wheel.
brakes	Almost all modern cars feature disc brakes all round. A few low-powered models still feature drum brakes at the back for cost reasons. "ABS" (antilock braking system) is increasingly fitted to all cars: it regulates brake application to prevent the brakes locking in an emergency or slippery conditions. "BA" (brake assist) is a system that applies maximum braking when it detects the driver hitting the brake pedal fast in an emergency, while "EBD" (electronic brake-force distribution) is a pressure regulator that spreads the car's braking torque more evenly so that the brakes do not lock. ESP, also known as ESC, VSC or DSC, is an electronically controlled program that helps keep the car pointing in the right direction in slippery conditions; sensors detect when the vehicle diverts from its intended course and apply the brakes asymmetrically to prevent skidding. "Brake-by-wire" is a totally electronic braking system that sends signals from brake pedal to brakes with no mechanical actuation whatsoever. TCS (traction-control system) is a feature that holds acceleration slip within acceptable levels to prevent wheelspin and therefore improves adhesion to the road.
tires	The size and type of wheels and tires are given in the internationally accepted formula. Representative examples include: 315/70R17, 235/50VR18, 225/50WR17, 235/40ZR18 and 225/40ZR18.

In all cases the number before the slash is the tire width in millimeters. The number after the slash is the height-to-width ratio of the tire section as a percentage. The letter R denotes radial construction. Letters preceding R are a guide to the tire's speed rating, denoting the maximum safe operating speed. H tires can be used at speeds up to 130 mph, V up to 150 mph, W up to 170 mph, and Y up to 190 mph. Z-rated tires are for speeds of 150 mph. Finally, the last number is the diameter of the wheel in inches.

wheelbase	The exact distance between the centre of the front wheel and the centre of the rear wheel.
track front/rear	The exact distance between the centres of the two front or the two rear tires, measured across the car on the ground.
curb weight	The amount a car weighs with a tank of fuel, all oils and coolants topped off, and all standard equipment but no occupants.
CO_2 emissions	Carbon dioxide emissions, which are a direct result of fuel consumption. CO_2 contributes to the atmospheric "greenhouse effect." Less than 100 g/km is a very low emission, 150 g/km is good, 300 g/km is bad. Recent European proposals have put forward a fleet average CO_2 emissions level of 120 g/km, with vehicles above this level attracting fines or penalties. "PZEV" (partial zero-emission vehicle) refers to a low-level emission standard that was created to allow flexibility on ZEV standards in California.

Other terms

A-, B-, C-, D-pillars	Vertical roof-support posts that form part of a car's bodywork. The A-pillar sits between windshield and front door, the B-pillar between front and rear doors, the C-pillar between rear doors and rear window, hatchback or wagon rear side windows, and the D-pillar (on a wagon) between rear side windows and tailgate. Confusingly, however, some designs refer to the central pillar between front and rear doors as a B-pillar where it faces the front door and a C-pillar where it faces the rear one.
all-wheel drive	A system delivering the appropriate amount of engine torque to each wheel via a propshaft and differentials, to ensure that tire slippage on the road surface is individually controlled. This system is ideal for high-performance road cars, such as Audis, where it is called "quattro."
beltline	*See* daylight opening line.
cant rail	The structural beam that runs along the tops of the doors.
coefficient of drag (Cd)	Shorthand for the complex scientific equation that proves how aerodynamic a car is. The Citroën C-Airdream, for example, has a Cd of 0.28, but the Citroën SM of thirty years ago measured just 0.24, so little has changed in this respect. "Drag" means the resistance of a body to airflow, and low drag means better penetration, less friction, and therefore more efficiency, although sometimes poor dynamic stability.
daylight opening line (DLO)	The line where the door glass meets the door panel, sometimes referred to as beltline or waistline.
diffuser	A custom-designed airflow conduit, often incorporated under the rear floor on high-performance and competition cars, which controls and evenly distributes fast-moving airflow out from beneath the speeding car. This ducting arrangement slows the flow of rushing air behind the car, lowering its pressure and so increasing aerodynamic downforce. The result is improved roadholding.
drive-by-wire technology	Increasingly featured on new cars, these systems do away with mechanical elements and replace them by wires transmitting electronic signals to activate such functions as throttle control, brakes, and steering.
fairing	A sculpted body surface blending different parts of a vehicle together to achieve a streamlined effect.
fast windshield	A windshield angled acutely to reduce wind resistance and accentuate a sporty look.

fastback	This refers to the profile of a hatchback that has a rear screen at a shallow angle, so that the tailgate forms a constant surface from the rear of the roof to the very tail end of the car.
feature line	A styling detail usually added to a design to increase interest and differentiate it from its rivals, and generally not related to such functional areas as door apertures.
flex-fuel	A flex-fuel vehicle (FFV) is an automobile that can accept a variety of different fuels, usually in the same tank. The most common current example is a vehicle that can run on blends of gasoline and bioethanol, ranging from 100% gasoline to 85% bioethanol–15% gasoline. Dual-fuel vehicles carry an additional natural gas tank and can switch between gasoline and natural gas.
four-wheel drive	This refers to a system delivering a car's power to its four wheels. In a typical "off-road"-type four-wheel-drive vehicle, the differentials can be locked so that all four wheels move in a forward direction even if the tires are losing grip with the road surface. This makes four-wheel drive useful when traveling across uneven terrain.
glasshouse/greenhouse	The car-design industry's informal term for the glazed area of the passenger compartment that usually sits above the car's waist level.
head-up display	A technology by which useful data are projected upward onto the inside of the windshield so that information can be displayed in the driver's line of vision.
high-intensity discharge (HID) headlamps	HID headlamps use an electric arc to produce the light, and are also known as xenon headlamps because of the gas used in the lamp. These lamps produce an immediate high-intensity light when switched on.
HVAC	"Heating, ventilation and air-conditioning system."
hybrid vehicle	A vehicle powered by a combination of two power sources, with a computer program continually determining the most efficient combination. The most familiar pairing, as on the Toyota Prius, is that of an electric motor for low speeds and a gasoline engine for faster driving. An on-board battery is able to store energy recuperated under braking (see regenerative braking). Other combinations, such as diesel/electric and fuel cell/electric, are also possible.
hydrogen fuel cell	A fuel cell produces electricity by combining hydrogen from an onboard tank with oxygen from the atmosphere. The only waste product is water, making fuel-cell vehicles emission-free at the point of use. Practical problems still to be overcome include cold-climate operation, hydrogen storage, and the cost of the proton exchange membrane (PEM) at the heart of the cell.
instrument panel	The trim panel that sits in front of the driver and front passenger.
Kamm tail	Sharply cut-off tail that gives the aerodynamic advantages of a much longer, tapering rear end, developed in racing in the 1960s.
monospace/ monovolume/"one-box"	A "box" is one of the major volumetric components of a car's architecture. In a traditional sedan, there are three boxes: one for the engine, one for the passengers, and one for the luggage. A hatchback, missing a trunk, is a "two-box" car, while a large MPV, such as the Renault Espace, is a "one-box" design, also known as a "monospace" or "monovolume."
MPV	Short for "multipurpose vehicle," this term is applied to tall, spacious cars that can carry at least five passengers, and often as many as nine, or versatile combinations of people and cargo. The 1983 Chrysler Voyager and 1984 Renault Espace were the first. The 1977 Matra Rancho was the very first "mini-MPV," but the 1991 Mitsubishi Space Runner was the first in the modern idiom. Commonly known in the USA as a minivan.
packaging space	Any three-dimensional zone in a vehicle that is occupied by component parts or used during operation of the vehicle.
platform, architecture	The assembly of elements that gives a car motive power and contact with the road: engine, gearbox, driveshaft, wheels, brakes, suspension, and steering. This assembly of components is loosely known as a platform, and, to save on development costs, a single platform can form the basis for several different models. Increasingly, designers refer to "architectures"—families that share basic building blocks, allowing different cars to be built on the same assembly line.

powertrain	The engine, gearbox, and transmission "package" of a car.
regenerative braking	When braking in a hybrid electric vehicle, the electric motor that is used to propel the car reverses its action and turns into a generator, converting kinetic energy into electrical power, which is then stored in the car's batteries.
rotary engine	The rotary engine is very different from a conventional piston engine, being, essentially, a triangular-sectioned shaft that rotates within an elongated chamber to create the compression and combustion cycle. It was developed by Felix Wankel in the 1950s.
shift paddles	A term used for steering-column-mounted levers that, when pulled, send electronic signals to the gearbox, requesting a gear change. They were first used in Formula One motor racing.
spaceframe	A structural frame that supports a car's mechanical systems and cosmetic panels.
splitter	Sometimes found at the front of high-performance cars near ground level, this is a system of undercar ducting that splits the airflow sucked under the car as it moves forward, so the appropriate volume of cooling air is distributed to both radiator and brakes. It also minimizes aerodynamic lift on the front axle.
subcompact	You need to rewind fifty-eight years for the origins: in 1950, Nash launched its Rambler, a two-door model smaller than other mainstream American sedans. The company coined the term "compact" for it, although, by European standards, it was still a large car. Nash's descendant American Motors then invented the "subcompact" class in 1970 with the AMC Gremlin, a model with a conventional hood and a sharply truncated hatchback tail; this was quickly followed by the similar Ford Pinto and Chevrolet Vega. In the international car industry today, "subcompact" is used as another term for "A-segment," the smallest range of cars, intended mostly for city driving.
SUV	Short for "sport utility vehicle," a four-wheel-drive car designed for leisure off-road driving but not necessarily agricultural or industrial use. Therefore a Land Rover Defender is not an SUV, while a Land Rover Freelander is. The line between the two is sometimes difficult to draw, and identifying a pioneer is tricky: SUVs as we know them today were defined by Jeep in 1986 with the Wrangler, Suzuki in 1988 with the Vitara, and Daihatsu in 1989 with the Sportrak. There is also a trend toward more sporty trucks, which has led to the more specific term "SUT," or "sport utility truck."
swage line	A groove or moulding employed on a flat surface to stiffen it against warping or vibration. In cars, swage lines add "creases" to bodywork surfaces, enabling designers to bring visual, essentially two-dimensional interest to body panels that might otherwise look slab-sided or barrel-like.
Targa	Porsche had been very successful in the Targa Florio road races in Sicily, so, in celebration, in 1965 the company applied the name "Targa" (the Italian for shield) to a new 911 model that featured a novel detachable roof panel. It is now standard terminology for the system, although a Porsche-registered trademark.
telematics	Any individual communication to a car from an outside base station, or vice versa; this could comprise, for example, satellite navigation signals, automatic emergency calls, roadside assistance, traffic information, and dynamic route guidance.
transaxle	Engineering shorthand for "transmission axle": this is the combined gearbox and differential unit that is connected to the driveshaft to transfer power to the driven wheels. All front-wheel-drive and rear- or mid-engined rear-wheel-drive cars have some type of transaxle.
tumblehome	The angle of the door glass when viewing a car from the front. The more upright the glass, the less tumblehome.
venturi tunnel	A venturi is an air-management system under a car designed to increase air speed by forcing it through tapered channels. High air speed creates a low-pressure area between the bottom of the car and the road, which in turn creates a suction effect holding the car to the road. Pressure is then equalized in the diffuser at the rear of the car.
waistline	*See* daylight opening line.

Where the New Models Were Launched

New York International Auto Show
April 6–15, 2007

Concept
Chevrolet Beat
Chevrolet Groove
Chevrolet Trax

Production
Ford Flex
Infiniti G37
Subaru Impreza

Seoul International Motor Show
April 6–15, 2007

Concept
Hyundai Veloster
Kia KND

Shanghai Motor Show
April 22–28, 2007

Concept
Audi Cross Coupé
BMW CS
Buick Riviera
Roewe W2

Frankfurt International Motor Show
September 13–23, 2007

Concept
Citroën C5 Airscape
Citroën C-Cactus
Ford Verve
Hyundai i-Blue
Kia Kee
Mercedes-Benz F700
Mitsubishi Concept-cX

Nissan Mixim
Opel/Vauxhall Flextreme
Peugeot 308 RC Z
Renault Kangoo Compact
  Concept
Renault Laguna Coupé
  Concept
SEAT Tribu
Volkswagen Up! & Space
  Up!

Production
Audi A4
BMW 1 Series Coupé
Fiat 500
Ford Kuga
Jaguar XF
Lamborghini Reventón
Mazda6
Mini Clubman
Opel/Vauxhall Agila
Peugeot 308
Renault Kangoo
Renault Laguna
Suzuki Splash
Volkswagen Tiguan

Tokyo Motor Show
October 27 – November 11, 2007

Concept
Audi Metroproject Quattro
Daihatsu Mud Master
Honda CR-Z
Honda Puyo
Lexus LF-Xh
Mazda Taiki
Mitsubishi Concept ZT
Nissan NV200
Nissan Pivo 2
Nissan Round Box
Subaru G4e

Suzuki PIXY & SSC
Suzuki X-Head
Toyota 1/X
Toyota Hi-CT
Toyota i-Real
Toyota RiN

Production
Nissan GT-R

Greater LA Auto Show
November 16–25, 2007

Concept
Audi Cross Cabriolet
  Quattro

Production
Honda FCX Clarity
Lexus LX 570
Lincoln MKS
Nissan Murano
Pontiac Vibe
Toyota Corolla
Toyota Matrix
Toyota Sequoia

North American International Auto Show (NAIAS)
January 19–27, 2008

Concept
Cadillac CTS Coupé
Cadillac Provoq
Chrysler ecoVoyager
Dodge Zeo
Ford Explorer America
Honda Pilot
Hummer HX
Jeep Renegade
Land Rover LRX
Lincoln MKT
Mazda Furai

Mercedes-Benz Vision
  GLK Freeside
Mitsubishi Concept RA
Nissan Forum
Saab 9-4X Biopower
Toyota A-BAT

Production
BMW X6
Dodge Ram
Fisker Karma
Ford F-150
Hyundai Genesis
Infiniti EX
Subaru Forester
Toyota Venza
Volkswagen Passat CC

Chicago Auto Show
February 8–17, 2008

Concept
GMC Denali XT

Production
Chevrolet Traverse
Dodge Challenger
Volkswagen Routan

Geneva International Motor Show
March 6–16, 2008

Concept
Fioravanti Hidra
Hyundai HED-5 i-Mode
Italdesign Quaranta
Kia Soul
Magna Steyr Mila Alpin
Opel/Vauxhall Meriva
Pininfarina Sintesi
Renault Megane Coupé
Saab 9-X BioHybrid
SEAT Bocanegra
Suzuki Concept A-Star

Production
Citroën C5
Dacia Sandero
Fiat Fiorino Panorama
Ford Fiesta
Honda Accord
Infiniti FX
Ken Okuyama K.07 & K.08
Lancia Delta
Renault Koleos
Skoda Superb
Tata Nano
Toyota iQ
Toyota Urban Cruiser
Volkswagen Scirocco
Volvo XC60
Zagato Bentley GTZ

New York International Auto Show
March 21–30, 2008

Concept
Kia Koup
Scion Hako

Production
Acura TSX
Honda Jazz/Fit
Hyundai Genesis Coupé
Nissan Maxima

Major International Motor Shows 2008–2009

Shows listed below cover the period November 2008 to October 2009.

Greater LA Auto Show
November 21–30, 2008
Los Angeles Convention Center, Los Angeles, USA
laautoshow.com

Essen Motor Show
November 29 – December 7, 2008
Messe Essen, Essen, Germany
essen-motorshow.de

Motor Show di Bologna
(Salone Internazionale dell'Automobile)
December 6–14, 2008
BolognaFiere, Bologna, Italy
motorshow.it

North American International Auto Show
(NAIAS)
January 17–25, 2009
COBO Center, Detroit, USA
naias.com

Geneva International Motor Show
March 5–15, 2009 (date to be confirmed)
Palexpo, Geneva, Switzerland
salon-auto.ch

Seoul International Motor Show
April 2–12, 2009
Kintex, Gyeonggi-do, Korea
motorshow.or.kr

New York International Auto Show
April 10–19, 2009
Jacob Javits Convention Center, New York, USA
autoshowny.com

Auto Shanghai
April 20–28, 2009
Shanghai New International Expo Center,
Shanghai, China
http://autoshanghai.auto-fairs.com/

Frankfurt International Motor Show
September 17–27, 2009
Trade Fairgrounds, Frankfurt am Main, Germany
iaa.de

Tokyo Motor Show
October 23 – November 8, 2009
Nippon Center, Makuhari, Chiba, Tokyo, Japan
tokyo-motorshow.com

Makes and Their Parent Companies

Hundreds of separate carmaking companies have consolidated over the past decade or so into major groups, including BMW, Chrysler, Daimler, Fiat, Ford, General Motors, Honda, Hyundai, Proton, PSA–Peugeot Citroën, Renault–Nissan, and Toyota. These account for at least nine out of every ten cars produced globally today. The remaining independent makes either produce specialist models, offer niche design and engineering services, or tend to be at risk because of their lack of economies of scale. The global overcapacity in the industry means that manufacturers are having to offer increased choice to the consumer to differentiate their brands and maintain market share. Not all parent companies fully own the carmakers listed as under their control. Mazda, for example, is a key part of the Ford Alliance but is only one-third owned by the US giant. Renault–Nissan, likewise, controls Russia's AvtoVAZ, even though (at time of going to press) it only holds 25 percent.

The past twelve months have seen major changes on the industrial front. Daimler and Chrysler split back into their constituent halves in autumn 2007; Toyota strengthened its control over Subaru in early 2008, when Renault–Nissan also took a stake in AvtoVAZ, maker of the Lada; Ford sold Aston Martin to a group of investors in spring 2007; the deal to sell Jaguar and Land Rover to India's Tata Motors was announced in March 2008; and Porsche now effectively controls the Volkswagen Group, having built up a stake and being committed to taking a major share (both companies come under a new entity, Porsche Holding).

BMW
BMW
Mini
Riley*
Rolls-Royce
Triumph*

Chrysler LLC
Chrysler
De Soto*
Dodge
Hudson*
Imperial*
Jeep
Nash*
Volvo

Daimler AG
Maybach
Mercedes-Benz
Smart

Fiat Auto
rth
meo
nchi*

Ford
Ford
Lincoln
Mazda
Mercury
Volvo

General Motors
Buick
Cadillac
Chevrolet
Corvette
Daewoo
GM
GMC
Holden
Hummer
Oldsmobile*
Opel
Pontiac
Saab
Saturn
Suzuki
Vauxhall

Honda
Acura
Honda

Hyundai
Asia Motors
Hyundai
Kia

Porsche Holding
Porsche
Volkswagen
  Group (see
  below)

Proton
Lotus
Proton

PSA–Peugeot Citroën
Citroën
Hillman*
Humber*
Panhard*
Peugeot
Simca*
Singer*
Sunbeam*
Talbot*

Renault–Nissan Alliance
Alpine*
AvtoVAZ (Lada)
Dacia
Datsun*
Infiniti
Nissan
Renault
Renault Sport
Samsung

SAIC (Shanghai Automotive International Co)
Austin
MG
Morris
Nanjing
Roewe
Ssang Yong
Wolseley

Tata Motors
Daimler
Indicar
Jaguar
Land Rover
Rover*
Tata

Toyota
Daihatsu
Lexus
Scion
Subaru
Toyota
Will*

Volkswagen Group
Audi
Auto Union*
Bentley
Bugatti
DKW*
Horch*
Lamborghini
NSU*
SEAT
Skoda
Volkswagen
Wanderer*

Independent makes
Aston Martin
  (Lagonda*)
Austin-Healey*
AviChina
Beijing
Bertone
Bristol
Byd
Caterham
Chery
Dongfeng
Donkervoort
EDAG
Elfin
ETUD
Farboud
Fenomenon
Fioravanti
Fisker Coachbuild
Gaz Group (LDV,
  Volga)
Heuliez
Hindustan
Inovo
Invicta
Irmscher
Isuzu
Italdesign
Izh

Jensen
Joss
Koenigsegg
KTM
Mahindra
Marcos
Maruti
Mitsubishi
Mitsuoka
Morgan
Pagani
Panoz
Paykan
Perodua
Pininfarina
Rinspeed
Sivax
Spyker
Stola
Th!nk
Tramontana
TVR*
Venturi
Westfield
Wiesmann
Zagato
ZAZ
ZIL

* Dormant makes

Acknowledgments

This book could not have been written without the help and support of a number of people. I would like to extend a special thank you to the team at Merrell Publishers for their professional work in helping to create another edition in the *Car Design Yearbook* series. Particular thanks go to Marion Moisy, Kirsty Seymour-Ure, Paul Shinn, John Grain and Michelle Draycott.

I would also like to pass on my thanks to Tony Lewin, who this year had a significant involvement acting as co-author and brought extensive knowledge of the industry to the editorial team, and to the manufacturers themselves for supplying information and photographs.

Stephen Newbury
Henley-on-Thames, Oxfordshire
2008

Picture Credits

MERRELL

First published 2008 by Merrell Publishers Limited

Head office
81 Southwark Street
London SE1 0HX

New York office
740 Broadway, Suite 1202
New York, NY 10003

merrellpublishers.com

A catalog record for this book is available from
the Library of Congress.

ISBN-13: 978-1-8589-4419-7
ISBN-10: 1-8589-4419-8

Project-managed by Marion Moisy
Copyedited by Kirsty Seymour-Ure
Proofread by Barbara Roby
Americanized by Chuck Brandstater
Designed by John Grain
Design concept by Kate Ward

Printed and bound in China

Frontispiece: Renault Mégane Coupé
Pages 6–7: Mercedes-Benz F700
Pages 16–17: Ford F-150
Pages 30–31: Audi A4
Pages 288–89: Peugeot 308 RC Z
Pages 302–303: Nissan GT-R